Institutional Dynamics

Earth System Governance: A Core Research Project of the International Human Dimensions Programme on Global Environmental Change
Frank Biermann and Oran R. Young, series editors

Institutional Dynamics: Emergent Patterns in International Environmental Governance
Oran R. Young

Related books from Institutional Dimensions of Global Environmental Change: A Core Research Project of the International Human Dimensions Programme on Global Environmental Change

Institutions and Environmental Change: Principal Findings, Applications, and Research Frontiers
Oran R. Young, Leslie A. King, and Heike Schroeder, editors

Managers of Global Change: The Influence of International Environmental Bureaucracies
Frank Biermann and Bernd Siebenhüner, editors

Institutional Dynamics

Emergent Patterns in International Environmental Governance

Oran R. Young

The MIT Press
Cambridge, Massachusetts
London, England

For information about special quantity discounts, please e-mail special_sales@mitpress.mit.edu.

This book was set in Sabon by Graphic Composition, Inc.
Printed and bound in the United States of America.

Library of Congress Cataloging-in-Publication Data

Young, Oran R.

Institutional dynamics : emergent patterns in international environmental governance / Oran R. Young.
p. cm. — (Earth system governance, a core research project of the international human dimensions programme on global environmental change)
Includes bibliographical references and index.
ISBN 978-0-262-01438-0 (hardcover : alk. paper) — ISBN 978-0-262-51440-8 (pbk. : alk. paper) 1. Environmental policy—International cooperation. 2. Environmental protection—International cooperation. I. Title.
JZ1324.Y68 2010
333.7—dc22

2009046142

10 9 8 7 6 5 4 3 2 1

To Linda and Jamie
The next generation in the quest for a better world

Contents

Series Foreword

Humans now influence all biological and physical systems of the planet. Almost no species, no land area, and no part of the oceans has remained unaffected by the expansion of the human species. Recent scientific findings suggest that the entire Earth system now operates outside the normal state exhibited over the past 500,000 years. Yet at the same time, it is apparent that the institutions, organizations, and mechanisms by which humans govern their relationship with the natural environment and global biogeochemical systems are utterly insufficient—and poorly understood. More fundamental and applied research is needed.

Yet such research is no easy undertaking. It must span the entire globe because only integrated global solutions can ensure a sustainable coevolution of natural and socioeconomic systems. But it must also draw on local experiences and insights. Research on Earth system governance must be about places in all their diversity, yet seek to integrate place-based research within a global understanding of the myriad human interactions with the Earth system. Eventually, the task is to develop integrated systems of governance, from the local to the global level, that ensure the sustainable development of the coupled socioecological system that the Earth has become.

The series Earth System Governance is designed to address this research challenge. Books in this series pursue this challenge from a variety of disciplinary perspectives, at different levels of governance, and with a plurality of methods. Yet all will further one common aim: analyzing current systems of Earth system governance with a view to increased understanding and possible improvements and reform. Books in this series will be of interest to the academic community but will also inform practitioners and at times contribute to policy debates.

This series is related to the long-term international research effort "Earth System Governance Project," a core project of the International Human Dimensions Programme on Global Environmental Change.

Frank Biermann, Vrije Universiteit Amsterdam
Oran R. Young, University of California, Santa Barbara
Earth System Governance series editors

Preface

This book is the latest product of a career-long interest in the roles that social institutions play in guiding the course of human-environment interactions. This time my focus is on institutional dynamics and especially what I call emergent patterns in environmental governance. My work in this field began in the 1970s with the introduction of a distinction between governance and government, a move meant to provide a point of departure for analyzing the supply of governance in anarchical settings (e.g., international society) lacking a government in the normal sense but often harboring an array of institutional arrangements relevant to the supply of governance. This move has proven fruitful in moving us beyond sterile debates on the need for some sort of international or world government. But because the specific institutions we call environmental or resource regimes, much like markets for specific products, are not material entities, we also face perennial challenges in efforts to identify actual regimes and to observe their operations in real-world settings. The facts that some regimes are not rooted in formal intergovernmental agreements and that the rules embedded in day-to-day practices often differ significantly from the rules on paper exacerbate these challenges.

During the 1970s and into the 1980s, I thought mainly about the formation or creation of regimes. Why do regimes form in response to some problems and in some settings but not in others? How can we account for the specific content of those regimes that do form? I sought to formulate a middle ground between the Hayekian trust in the development of spontaneous or self-generating governance systems (Hayek 1973) and the realist belief that social institutions are largely epiphenomena reflecting the underlying distribution of power in social settings (Mearsheimer 1994/1995). This led to an extended effort to understand what I call institutional bargaining and to identify those features that differentiate institutional bargaining from other sorts of bargaining (Young 1994).

By the 1990s, my attention had shifted to the effectiveness of regimes, seeking answers to questions about how, when, and why regimes matter in the sense of influencing the course of human-environment interactions. These issues, it turns out, are not only matters of fundamental importance with regard to the idea of governance but are also unusually challenging in analytical and methodological terms. Here, too, I have sought to devise a middle way between those who dismiss regimes—and social institutions more generally—as mere surface phenomena having little or no causal significance (Mearsheimer 1994/1995) and those who believe that institutions constitute the dominant driver of the course of human-environment interactions (North 1990). A factor that has made the study of regime effectiveness particularly challenging as well as endlessly intriguing is the central role of complex causality in this domain. Regimes do make a difference in determining the course of human-environment interactions. But so do a variety of other biophysical factors (e.g., climate and weather patterns) and socioeconomic forces (e.g., technology). To make matters even more challenging, many of these factors interact with one another, giving rise to causal complexes. The trick is to devise ways to understand how these complexes drive human-environment interactions in settings where it is often impossible to reach conclusions about the proportion of the variance in some well-defined dependent variable we can attribute with confidence to individual and well-behaved drivers.

The effort to understand effectiveness has consumed a sizable proportion of the time and energy I have been able to devote to research for many years; it does so still. But, all along, I have been aware that environmental and resource regimes are dynamic systems. It does not take much contact with the real world to realize that these institutional arrangements do not remain unchanged following their initial creation. Most regimes change continuously during the course of their lives. Some undergo dramatic changes or what my colleagues and I called "watershed changes" in a sustained effort to study regimes quantitatively starting in the 1990s (Breitmeier, Young, and Zürn 2006). But no regime is static in the sense that it remains completely unchanged over the course of time. I have been meaning to devote some systematic thought to the phenomenon of institutional change for some time. This book is the product of that effort.

There are many questions about institutional change that I do not address in this study. I focus on emergent patterns in environmental governance arising from the operation of regimes during the period following their creation. This leads to an effort to step back from the day-to-day operations of regimes, to look at their performance through time, and to characterize the big picture that emerges from this examination. I do not engage in a more microscopic or blow-by-blow analysis of the occurrence of specific changes. This leaves ample room for others who develop an interest in institutional dynamics to contribute to this field of study. My goal in the work that follows is to take one step toward the development of a useful and usable theory of institutional dynamics.

I have been fortunate to be able to participate actively and more or less simultaneously in three distinct communities while working on this book. The scientific community encompassing those endeavoring to generate new knowledge about international institutions has the virtue of embracing and applying rigorous methodological standards to work in this field. The global-change research community directs attention to finding ways to understand complex and dynamic systems whose behavior is affected by interactions among a range of biophysical and socioeconomic forces. The community of practitioners—those responsible for creating and administering regimes dealing with specific issues like climate change—is concerned with policy relevance and has a low tolerance for discussions about methodology or debates about theoretical matters of interest to scholars (e.g., the relative importance of power, interests, and knowledge as determinants of the course of institutional change). It is a challenge to be able to operate in all three communities in a constructive and productive manner. I cannot judge how successful my efforts to do so will seem to others. But I do know that the quality of my understanding of emergent patterns in environmental governance rests squarely on insights that have come my way as a result of operating in all three communities at the same time.

I have been blessed during my years at the Bren School of Environmental Science and Management at UC Santa Barbara with excellent colleagues, eager students, and a superb assistant. Durwood Zaelke and Matthew Stilwell, with whom I have sought to explore the idea of governance for sustainable development, have become good friends as well as excellent colleagues. My students, both at the master's level and at the PhD level, have been a source of continuous inspiration in an organizational setting that has sometimes proved frustrating. They have taken up the challenge of seeking to understand the nature and impacts of environmental and resource regimes with both eagerness and a sense of commitment to exploring how social institutions affect our lives and to learning how to create and administer effective regimes. Maria Gordon, my assistant, has been a tower of strength throughout. There is simply no way that I could stay afloat in the sea of obligations I have acquired without her able and amiable assistance. Three anonymous reviewers for the MIT Press not only drew my attention to conceptual matters in need of clarification; they also saved me from a sizable number of small but significant errors in my treatment of individual regimes. They performed their designated role admirably. To all these people, I want to take this opportunity to say thanks and to convey my gratitude for the roles they play in ensuring that there is never a dull moment in the life I lead.

1

Emergent Patterns: Concepts and Hypotheses

Introduction

Like all social institutions, environmental and resource regimes—assemblages of rights, rules, and decision-making procedures that influence the course of human-environment interactions—are dynamic. Some changes, such as the adoption of significant amendments to an existing statute or the addition of substantive protocols to a framework convention, are developmental in character. They serve to flesh out or update existing regimes, to extend the coverage of these arrangements to new issue areas, and generally to enhance the effectiveness of these governance systems. Other changes are better understood as responses to external events involving the biophysical, socioeconomic, or technological settings in which regimes operate. They are responses designed to maintain the resilience of regimes in changing settings. Some changes are gradual and incremental in nature; they develop step-by-step over relatively long periods and often take the form of informal adjustments in the practices that grow around specific regimes. Others are more abrupt and far-reaching; they can and sometimes do involve state changes in regimes treated as complex systems and may precipitate major restructuring (or even replacement) of formal constitutive agreements.

In one form or another, change is a pervasive feature of the landscape of environmental and resource regimes. All regimes experience change on a continuous basis. But can we discern emergent patterns or capture key elements of the overall picture of change arising over time with regard to individual regimes? Do some regimes go from strength to strength, building on early successes to become more and more effective in solving problems with the passage of time? Do others run into roadblocks early on that stymie their capacity to solve problems? Are there cases in which highly effective regimes run out of steam and fall into a phase of decline and, conversely, cases in which regimes that are ineffective at first turn a corner and become substantially more successful in later stages?

Understanding these emergent patterns in environmental governance is important for efforts to solve many problems associated with human-environment interactions (e.g., climate change, the loss of biological diversity, or the degradation of marine systems). The need to understand these patterns is destined to become increasingly prominent in the foreseeable future. Yet the literature on this topic is thin, and our knowledge of institutional dynamics is comparatively underdeveloped (Keohane and Nye 1977; Young 1999a: ch. 6). This is particularly true regarding multilateral environmental agreements (MEAs) and other forms of international and transnational cooperation pertaining to environmental issues. But knowledge of the dynamics of environmental and resource regimes operating at lower levels on the scale of social organization is not much more advanced. The goal of this book is to take some initial steps toward filling this gap in knowledge.

The sources of this gap are easy to identify. During the 1980s and into the 1990s, the research community focused on conceptual questions (e.g., how to define the term *regime* or how to identify individual regimes empirically) and directed attention to issues pertaining to regime formation to explain when and why efforts to form regimes designed to deal with specific problems succeed or fail (Young 1982; Krasner 1983b; Keohane 1984; Haggard and Simmons 1987; Young 1989; Ostrom 1990; Rittberger and Mayer 1993; Young and Osherenko 1993; Young 1994; Hasenclever, Mayer, and Rittberger 1997). Because this line of analysis called for a shift in focus from organizations to institutions, it made sense to bear down hard on clarifying the unit of analysis and delineating the universe of cases.

During the 1990s and on into the opening years of this decade, we directed more and more attention to matters of regime effectiveness and performance (Haas, Keohane, and Levy 1993; Levy, Young, and Zürn 1995; Young 1999b; Miles et al. 2002; Breitmeier, Young, and Zürn 2006). Because prominent critics cast doubt on the causal significance of these social institutions (Strange 1983; Mearsheimer 1994/1995) and because the problems involved in demonstrating causality in this realm are difficult to solve, we have devoted the lion's share of our time and energy to devising methods for overcoming or circumventing these analytic obstacles in the interests of illuminating various aspects of regime effectiveness (Underdal and Young 2004; Young et al. 2006b). There is nothing surprising or inappropriate about these allocations of intellectual resources. But this understandable intellectual history must not be allowed to deflect attention from efforts to make progress in the analysis of institutional dynamics during the next phase of research on regimes.

My purpose in this introductory chapter is to set the stage for a productive encounter between analytical perspectives and empirical observations derived from a series of in-depth case studies. I approach the effort to understand patterns of institutional change as a study of the behavior of complex and dynamic systems. I begin with a consideration of the recent literature on resilience, vulnerability, and adaptation in socioecological systems (Berkes and Folke 1998; Gunderson and Holling 2002; Walker and Salt 2006; Young et al. 2006a) and an assessment of the value of bringing this analytic framework to bear on the subject of institutional change.

The chapter proceeds as follows. The first substantive section sets out the conceptual framework for the analysis to come. It develops the case for treating regimes as complex and dynamic systems and discusses the applicability of a number of concepts commonly used in analyzing socioecological systems (e.g., robustness, resilience, stress, or state change) to institutions construed as systems. The next section focuses on emergent patterns in environmental governance. It starts with an account of the idea of a pattern of change and goes on to introduce the five major emergent patterns to be examined through in-depth case studies in the chapters to come. The following section sets forth an argument about the determinants of emergent patterns of change under real-world conditions. It introduces what I call the endogenous-exogenous alignment thesis that serves as the principal theoretical proposition to be examined systematically in the case studies presented in chapters 2 through 6 and revisited in the book's concluding chapter. The final substantive section provides brief introductions to the cases I have chosen for in-depth analysis in an effort to deepen our understanding of emergent patterns in environmental governance. The chapter's conclusion opens the door to the empirical heart of the book.

The cases I examine in depth in search of the determinants of patterns of change deal with large-scale terrestrial, marine, and atmospheric issues. But interest in patterns of institutional change is generic. The argument applies to the full range of environmental and resource regimes operating at all levels of social organization; it may be of interest to those interested in exploring the dynamics of social institutions operating in other issue areas as well.

Regimes as Complex and Dynamic Systems

The starting point for this inquiry is the proposition that it is fruitful to approach institutional dynamics through the lens of complex and dynamic systems. Most research on such systems focuses on ecosystems, social systems, and, increasingly, socioecological systems. But it is not a stretch to apply the same conceptual lens to social institutions. Environmental and resource regimes—from the local level on up to the global—are systems in the sense that they are made up of interconnected elements (e.g., rights, rules, and decision-making procedures) that are organized around some function or purpose (e.g., ensuring that humans use natural resources in a sustainable manner) and that are differentiable from the environments or settings (which may include other institutions) in which they operate (Meadows 2008).

As is the case with ecosystems and especially socioecological systems, the specification of both spatial and temporal boundaries separating specific institutions from their environments requires analysts to make judgments that go well beyond simple descriptive accounts (Levy, Young, and Zürn 1995). Regimes located at any given level of social organization typically are nested into institutional arrangements operating at higher levels that affect their implementation. The various regional seas regimes as well as many regional fisheries regimes, for example, all fit into the overarching governance system for marine issues articulated in the 1982 UN Convention on the Law of the Sea (UNCLOS). Such relationships require careful consideration. But there is no reason to allow this concern to cripple empirical research pertaining to formation, effectiveness, and change in specific environmental and resource regimes.

Treating regimes as complex and dynamic systems leads directly to the introduction of concepts like robustness, resilience, vulnerability, stress, and state change or regime shift as a vocabulary for asking and answering questions pertaining to institutional dynamics. It also makes it possible to link the analysis of institutional change to a large body of intellectual capital accumulated by observers concerned mainly with the behavior of socioecological systems (Gunderson and Holling 2002; Walker and Salt 2006).

Robustness is the capacity of a system to cope effectively with challenges and stresses without undergoing significant changes in its own elements or procedures (Anderies, Janssen, and Ostrom 2004). Arrangements that have built-in countercyclical mechanisms (e.g., predator-prey relationships in ecosystems or the interplay of supply and demand in competitive markets) are often robust in this sense. Regimes governing human uses of living resources, for instance, typically feature procedures for setting allowable harvest levels on an annual basis and for tracking closely—or even anticipating—shifts in the abundance of such resources relative to aggregate demand. Resilience, by contrast, refers to the "capacity of a system to experience disturbance and still maintain its ongoing functions and controls" (Holling and Gunderson 2002: 50).[1] Sometimes labeled "ecosystem resilience," this concept encompasses situations in which systems undergo significant changes in processes needed to cope with challenges and stresses without experiencing drastic change. Resilience thus includes variability as well as persistence in contrast to the idea of "engineering resilience," which starts from the concept of equilibrium and directs attention to the derivation of stability conditions specifying how far a system can be displaced from a fixed point of equilibrium and still return to that equilibrium once the disturbance has passed. Although engineering resilience is more familiar to most of us—not to mention easier to model—there is a compelling case to be made for the proposition that adopting the idea of ecosystem resilience as a point of departure makes sense for studies of institutional dynamics as well as the dynamics of socioecological systems (Holling and Gunderson 2002: 27–30).

An observation about the resilience of complex and dynamic systems that deserves emphasis at the outset is that there is nothing static about this way of thinking. Instead of returning to some well-defined equilibrium state (e.g., a specified temperature in a thermostatically controlled heating system or a fixed speed in a car equipped with cruise control), institutions—like socioecological systems—often develop or evolve in the sense that they move toward a realization of their potential or make adjustments needed to match changing biophysical or socioeconomic circumstances. The law of the sea as codified in UNCLOS, for instance, has evolved significantly through the addition of the straddling stocks agreement pertaining to fisheries cutting across jurisdictional boundaries, new rules pertaining to marine transport, regional seas arrangements focusing on the control of pollution, and so forth (Hoel 2000; Stokke 2001; Ebbin, Hoel, and Sydnes 2005). But no one would claim that these developmental processes have transformed the law of the sea or even, to use a phrase common in analyses of ecosystems, triggered a state change regarding the condition of the overall institutional arrangements governing human uses of marine resources.

Although there is some lack of conceptual consensus in the literature, vulnerability is a closely related concept (Adger 2006). Vulnerability rises as stresses begin to overwhelm a system's robustness (i.e., its capacity to handle stress without adapting) and challenge its resilience (i.e., its capacity to deal with stress through adaptation). While robustness and resilience refer to a system's capacity to cope with stress, vulnerability is a matter of sensitivity to stresses, disturbances, and threats of one sort or another. A system that is highly robust or resilient will be relatively invulnerable in the sense that it will be immune to the impacts of many disturbances. But there is no reason to assume that a system's robustness, resilience, and vulnerability will be uniform either across the full range of actual or potential stresses or from one time period to the next (Eggertsson 2005). An institution may be robust in the sense that it is more or less immune to the impact of most stresses, yet be highly vulnerable to one or more specific types of disturbance. The development of new technologies (e.g., the introduction of high-endurance stern trawlers in industrial fisheries during the 1970s and 1980s), for instance, can quickly undermine an institutional arrangement or governance system that has worked reasonably well over an extended period (Warner 1983). This is one reason why regimes that appear to most observers to be highly resilient or deeply entrenched can collapse quite suddenly and in a manner generally unanticipated by subjects and outside observers alike.

Within this frame of reference, the term *stress* refers to any force or process that increases vulnerability or degrades the robustness or resilience of a given system.[2] But what is stressful to one system may not disturb another system at all. Small-scale societies heavily dependent on the success of local subsistence or artisanal fishing, for example, are far more vulnerable to severe depletions of key stocks than counterparts that are able to shift back and forth among a number of food sources and obtain various goods and services through trade with outsiders. Conversely, small communities that have become heavily dependent on the international market for their products (e.g., coffee) are more vulnerable to external forces than others that have remained more self-sufficient. The sources of stress may be internal, external, or both. Societies may develop management regimes for fisheries that are capable (or incapable) of restricting harvest levels in a manner compatible with the sustainable use of the resource. Biophysical changes (e.g., El Niño/Southern Oscillation events) affecting the relevant systems may increase or decrease stress on an environmental or resource regime over and above problems arising from the operation of the regime itself. The simultaneous occurrence of several distinct threats or stresses is common. The forces producing institutional changes are particularly strong under such circumstances.

Dealing with a single atmospheric pollutant is one thing. But devising regimes capable of coping with the impacts of atmospheric emissions of sulfur dioxide, nitrogen oxides, volatile organic compounds, ozone-depleting substances, persistent organic pollutants, and greenhouse gases at the same time is far more challenging.

From this perspective, it is essential to consider mechanisms—arising spontaneously or created intentionally—available to cope with stress and even to take advantage of stresses to stimulate institutional development or reform. With regard to institutions, in particular, it is worth emphasizing right away that regimes vary in terms of their ability to create and implement new mechanisms to cope with stresses as the need arises. Regimes that are difficult to amend formally or to adjust informally run the risk of falling prey to rapid changes in either biophysical or socioeconomic conditions. Yet regimes that are too easy to alter become ineffective in terms of influencing the behavior of those whose actions have given rise to the problems these institutional arrangements are created to solve. A fisheries regime that lacks the capacity to reduce allowable harvest levels in a timely manner when key stocks decline, for instance, cannot play an effective role in addressing problems of fisheries management. But a regime that yields to the slightest sign of stress cannot be effective in guiding the behavior of those subject to its provisions. The challenge of striking a balance between extreme rigidity and excessive flexibility looms large in any effort to understand the nature of institutional dynamics. The regime created to protect stratospheric ozone, which draws a clear distinction between adjusting the rules pertaining to families of substances that are already regulated and adding new families of chemicals to the roster of regulated substances, offers a particularly interesting example of this sort of balancing.

More drastic changes occur when stresses overwhelm the capacity of a system's stress management mechanisms to cope with such pressures. Individual regimes vary enormously in these terms. Some experience high stress but also develop effective mechanisms to manage stress. Others are low-stress systems that do not need a highly developed capacity to manage stress. Real-world arrangements are often more complex than these general propositions suggest. Fisheries regimes that are generally robust or resilient can collapse in short order in the wake of seemingly small changes in biophysical conditions (e.g., slight variations in water temperatures or patterns of upwelling). Arrangements that are crisis prone in the sense that they have a limited capacity to cope with day-to-day stresses (e.g., the need to reduce allowable harvest levels in light of fluctuations in the condition of key stocks, or pressures to create more allowances in regimes governing emissions of greenhouse gases) can emerge as effective arrangements once the occurrence of a crisis is acknowledged. A major issue facing those responsible for operating specific regimes concerns the relative emphasis to place on anticipatory measures, or, in other words, efforts to build capacity to cope with stresses before they occur in contrast to adaptive measures, or, in still other words, actions that come into focus only after the pressures arising from stress become severe.

Emergent Patterns in Environmental Governance

Patterns of institutional change are emergent properties arising over time from the dynamics of complex systems. Only rarely are they products of intentional public choices, much less deliberative actions on the part of goal-oriented actors (Hayek 1973). But this does not mean that all regimes are unique in these terms, so that we should not expect to find distinct patterns of change in examining the dynamics of a range of environmental and resource regimes over time. The main goal of this book is to identify emergent patterns in real-world cases and to develop an analysis that can account for their occurrence.

I use the concept of a pattern of change to refer to the principal features or dominant elements of the experiences of environmental and resource regimes emerging through time. Regimes change all the time. But many individual changes have little or no significance in establishing broad patterns that emerge through time. In identifying emergent patterns of change, therefore, we need to separate the signal of patterns from the noise of a multitude of specific changes occurring continuously. There are cases in which this is hard to do, especially in the early days of regimes when broad patterns have yet to emerge. There is always an element of judgment in identifying emergent patterns of change in individual regimes; some disagreement regarding the characterization of these patterns is to be expected. Yet it is surprising how easy it is to identify patterns of this sort in most cases and how much consensus there is in the conclusions of those who think about such matters.

In my long-term research on environmental regimes (Young 1999a; Young 2002; Young, King, and Schroeder 2008), I have observed five distinct emergent patterns of change. I call them progressive development, punctuated equilibrium, arrested development, diversion, and collapse. These distinctions are analytical in character; not all cases of institutional change in the real world fit easily or comfortably into one of these patterns. Even in analytical terms, the five categories do not add up to a proper taxonomy in the sense of constituting a set of mutually exclusive and exhaustive categories. Other patterns may occur under specific conditions or come into focus as emergent patterns in the future. Yet the five patterns of change I examine in this book all occur in the real world with some frequency. A study of these patterns provides a way forward for those interested in adding to our understanding of the dynamics of environmental and resource regimes.

Progressive Development

Some regimes start from a well-defined, albeit sometimes modest, initial state and advance steadily in a manner that does not feature major challenges or severe setbacks and in a manner that increases the capacity of the regimes in question to address the problems they are created to solve. One path that exemplifies this pattern starts with a framework convention followed in fairly short order by one or more substantive protocols that are then amended and extended to accommodate new information about the nature of the problem or to build the capacity of the regime to influence the behavior of those who are subject to its assemblage of rights, rules, and decision-making procedures. But this is not the only trajectory featuring a process of institutional change that deserves the label progressive development. Even at the international level, some agreements create umbrella arrangements intended to provide overarching constitutive systems within which to tackle specific problems through the establishment of more focused substantive arrangements. The law of the sea articulated in the provisions of UNCLOS provides a clear example. The 1982 convention comprises a series of broad arrangements covering issues like Exclusive Economic Zones, the Area, and the high seas together with general provisions pertaining to activities like fishing, shipping, and environmental protection. More often than not, these broad provisions must be supplemented by more detailed arrangements covering particular resources or activities (e.g., the straddling stocks agreement regarding fisheries, or the regional seas arrangements regarding pollution) to give them bite in behavioral terms. There is no implication here that a large proportion of regimes will exhibit some form of progressive development. Progress of this sort is the exception rather than the rule in many issue areas. Nor is there any basis for assuming that progressive development will go on indefinitely, even in cases where this sort of development is clearly in evidence during various stages of institutional maturation. But there are cases of progressive development in the universe of environmental and resource regimes, and it is an important challenge to determine why this pattern occurs in some cases but not in others.

Punctuated Equilibrium

The pattern I call punctuated equilibrium differs from the pattern of progressive development in several respects. Unlike the steady advance that is the hallmark of progressive development, punctuated equilibrium occurs in cases where regimes encounter periodic stresses that challenge their capacity to adjust while also triggering episodes of regime building that are progressive in nature. This pattern may involve ecological challenges (e.g., changes in climatic conditions threaten the viability of living resources), economic challenges (e.g., corporations wish to initiate activities likely to have harmful environmental impacts), or political challenges (e.g., actors outside the regime begin to lobby for admission). The responses may vary from using existing procedures to cope with stresses to agreeing to add new members, negotiating supplemental protocols, or creating closely related arrangements covering emerging issues. Rising interdependencies in socioecological systems lead increasingly to the emergence of what astute observers call institutional complexes (Raustiala and Victor 2004) and to efforts to manage interactions among the components of the resultant metaregimes. Punctuated equilibrium thus encompasses a range of situations that may seem superficially to differ from one another in significant ways. But the common element in all these cases is the combination of recurrent challenges to the capacity of regimes to operate effectively and the adjustment of existing institutional arrangements or the creation of supplemental arrangements that allow regimes to meet new challenges, even as they continue to deal effectively with the problems that led to their creation in the first place.

Arrested Development

A somewhat related pattern features what I call arrested development. Here, regimes get off to a promising start but then run into barriers or obstacles that block further development. Such a pattern is relatively common with regard to framework conventions that fail to live up to their promise regarding the development of substantive protocols or the adjustment of existing protocols to maintain their relevance in the face of important changes in the nature of the problems they address. The framework-protocol example relates primarily to international regimes. But this basic pattern is common at the national level as well. Regimes that start with the passage of promising legislation amid considerable fanfare can get lost during the process of implementation and languish in political limbo at least until it comes time to reauthorize them and to update their provisions to deal effectively with changing circumstances. Cases of arrested development may be hard to distinguish from cases of progressive development at the point of inception. Whereas progressive development features arrangements that go from strength to strength and find ways to adjust to changing circumstances, cases of arrested development get stuck at some (usually early) stage in the developmental process and fail to overcome barriers blocking the path toward further development. We need to be careful in applying this analytical construct to real-world situations. Regimes that appear to be stuck in a political logjam can break through to a new era of progressive development; regimes that appear to be success stories can run into unexpected obstacles that arrest or severely hamper continued development. But the challenge is clear. We need to probe the record in search of factors that explain why some cases exhibit the pattern of progressive development, while others fall prey to the problem of arrested development.

Diversion

This pattern encompasses environmental and resource regimes that are created for one purpose but later are redirected in a manner that runs counter to the original purpose. Such developments may reflect the growth of knowledge about relevant biophysical systems (e.g., new insights regarding the importance of habitat preservation to the protection of species) or shifts in the normative outlooks or value systems of key players (e.g., the rise of preservation in contrast to conservation as a normative goal). Once diversion occurs, a regime may evolve to embrace some new agenda. But diversion is equally likely to be followed by a period of gridlock in which the regime has no clear direction. An interesting question in such cases centers on the motivations of the key players. Why not simply terminate an outmoded regime (e.g., the regime dealing with trade in endangered species) and replace it with new arrangements that are better suited to shifting circumstances (e.g., regimes that direct attention to the loss of habitat and the problem of invasive species as well as trade in endangered species). The answer to this question has two components. Existing regimes often prove sticky. It is not so easy to get rid of a regime, even when awareness of its positive and normative limitations becomes widespread. New regimes are hard to create, at least in part because they exhibit the attributes of public goods. There is no guarantee that new regimes will prove effective, even when those who create them are well-versed in new knowledge and accept the tenets of new paradigms. Regimes almost always affect the interests of stakeholders asymmetrically. It is to be expected that institutional arrangements will become arenas for ongoing struggles over who benefits from their operation and who suffers either from the consequences they generate or from the necessity of bearing a share of the burden of operating the regimes on a day-to-day basis. The pattern I call diversion applies to cases featuring state changes in ongoing regimes in contrast to the pulling and hauling that is a feature of everyday political processes within generally stable institutional frameworks. A shift within an existing regime from efforts to achieve maximum sustainable yields (MSY) from a harvested species to a policy of preservation in which all (intentional) killing is put off-limits counts as a diversion (Larkin 1977). A change involving the adjustment of regulations pertaining to the treatment of by-catches in specific fisheries does not qualify for this designation.

Collapse

The hallmark of my final pattern is the collapse of an institutional arrangement in the sense that it crosses some threshold leading relatively shortly either to the termination of the regime in formal terms or to a severe decline that transforms the regime into a dead letter, even though it continues to exist on paper. This pattern does not include cases in which regimes are stillborn and never give rise to practices that make a difference in terms of problem solving. Rather, collapse occurs when a regime has been in operation for some time but then encounters either endogenous shocks (e.g., allowable harvest levels that are set too high trigger the collapse of a fishery) or exogenous shocks (e.g., a convention is suspended because two or more of the parties to the agreement go to war with one another) that put an end to the arrangement in relatively short order. Collapse in this sense can occur at various stages in the development of a regime. An arrangement that fails to make progress toward solving the relevant problem can collapse early in the developmental sequence. But regimes that have performed well, sometimes for long periods, are also subject to collapse when a long-standing problem takes a turn for the worse, when new technology transforms the problem, or when the intellectual capital on which the arrangement is founded erodes and is replaced by a new paradigm. The advent of high-endurance stern trawlers during the 1970s and 1980s transformed many fisheries in ways that simply overwhelmed the capacity of long-standing regimes to adapt (Warner 1983; Harris 1998; Pauly et al. 1998). Similarly, the shift from thinking guided by the idea of MSY to a new paradigm centered on the idea of ecosystem-based management has changed the cognitive landscape in ways that a number of existing regimes have been unable to accommodate in the absence of drastic changes (Larkin 1977).

The Endogenous-Exogenous Alignment Thesis

Assuming these five patterns of institutional change occur with some frequency under real-world conditions, what are the causal mechanisms that determine which pattern will emerge in any particular case? Why has the regime created to protect stratospheric ozone become a case of progressive development, while the climate regime, at least at this stage, is properly characterized as a case of arrested development? What explains the ability of the Antarctic Treaty System to cope with serious challenges and to come through these episodes not only unscathed but also strengthened? What is the proper explanation for the diversion of the regime for whales and whaling from a conservation arrangement intended to make sustainable harvests possible to a regime dedicated to the preservation of whales? How can we account for the sudden collapse of the fur seal regime, an arrangement widely regarded in earlier times as an exemplar of success in the field of wildlife conservation?

Many observers have sought to answer such questions with regard to individual regimes; the resultant explanations tend to be ad hoc in nature and to privilege factors of interest to individual observers. We hear a lot about the limited economic significance of chlorofluorocarbons (CFCs) and halons in contrast to the industrial, commercial, and agricultural interests engaged in activities that generate emissions of greenhouse gases. It is common to make reference to shifts in prevailing discourses or paradigms in efforts to explain the collapse of the fur seal regime and the diversion of the regime for whales, both developments occurring during the 1980s. Many attribute the success of the Antarctic Treaty System to a combination of limited interest on the part of influential states and vigorous campaigns on the part of environmental groups like Greenpeace or the Antarctic and Southern Ocean Coalition.

Yet all these explanations seem limited in nature and, in the final analysis, unsatisfactory; they certainly do not add up to a theory of institutional change that can help us explain or even predict patterns of change occurring across the universe of international environmental and resource regimes. In the body of this book, encompassing a series of in-depth case studies followed by a concluding chapter designed to gather up the threads of the empirical analyses, I seek to take a step forward in understanding the determinants of patterns of change. Specifically, I develop and explore empirically the proposition that the patterns of change occurring in individual regimes are determined by interactions between endogenous, or regime-specific, factors and exogenous factors, or forces operative in the biophysical and socioeconomic settings in which regimes are located. I call this the endogenous-exogenous alignment thesis; it is the central theme of the book.

Endogenous factors are those having to do with attributes of the regimes themselves. They encompass a range of considerations such as the locus of the regime on a hard law–soft law continuum; the nature of the relevant decision rule(s); provisions for monitoring, reporting, and verification; funding mechanisms; procedures for amending a regime's assemblage of rights, rules, and decision-making procedures; and so forth. Exogenous factors include conditions pertaining to the character of the overarching political setting; the nature of the prevailing economic system; the rise of new actors, technological innovations, and the emergence of altered or entirely new discourses; as well as significant changes in broader biophysical systems. Neither of these categories of factors is strictly limited. There is always the prospect that one or more previously unidentified factors will emerge and play an important role in individual cases. Table 1.1 lists the major factors in each category. I refer to these factors on a regular basis in the case studies examined in the following chapters and present an assessment of the robustness of the endogenous-exogenous alignment thesis in the book's concluding chapter.

Patterns of change occurring in specific cases arise from interactions between these two sets of factors. When abrupt changes or exogenous shocks are uncommon and a regime is resilient in the sense that it has a good deal of adaptive capacity, for instance, the prospects for progressive development are good. Shifts in the character of the problem or increases in its severity coupled with institutional rigidity are likely to eventuate in arrested development or even collapse. The rise of a new paradigm (e.g., ecosystem-based management) in a broader issue area can undermine a regime that lacks a mechanism for shedding one paradigm and embracing another. Changes in the interests of key players can lead to the use of a regime's decision rules to alter the basic purposes of the regime, or, in my terminology, to divert the regime without forcing it out of existence. Exogenous forces can reinforce the status quo or provide the impetus for institutional change. The outcomes I focus on in analyzing emergent

Table 1.1
Determinants of Patterns of Regime Change

Endogenous factors	Exogenous factors
Locus on the hard law–soft law continuum	Attributes of the problem
Decision rules	Political (dis)continuity
Flexibility in the face of changing circumstances	Economic (in)stability
Monitoring, reporting, and verification procedures	Technological innovations
Administrative capacity	Emergence of new actors in the issue area
Resources/funding mechanisms	Shifting paradigms or discourses
Amendment procedures	State changes in relevant biophysical systems
	Exogenous shocks (e.g., the ozone hole)

patterns center on the roles regimes play in addressing the problems that lead to their creation. Is the seasonal thinning of stratospheric ozone becoming more or less severe? Are we taking steps to stabilize the climate system? Are we using renewable resources in a sustainable fashion? One immediate implication of this analytic perspective is that what is known in regime analysis as the problem of fit, or the compatibility (or incompatibility) of institutions and the broader settings or environments in which they operate, will be an important source of insights in efforts to account for patterns of institutional change (Young 2002; Young, King, and Schroeder 2008).

In virtually every case, complex causality is a salient feature of the institutional landscape. Single-factor analyses rarely suffice to satisfy our desire for answers to questions about emergent patterns in environmental governance. Perhaps for this reason, identifiable individuals often become key players, using their entrepreneurial skills to broker deals, articulating new ways of thinking about the problem, or bringing to bear the structural power or influence of states or nonstate actors on specific issues (Young 1991; Young and Osherenko 1993). Complexity creates a setting that is susceptible to the influence of those who are able to identify and proceed nimbly to take advantage of attractive openings.

We must go beyond sound bites in our efforts to understand patterns of institutional change; inquiries regarding combinations of interactive drivers will be needed to develop satisfactory explanations of what happens in specific cases (Young 2006b). There is no reason to regard this conclusion as disappointing or discouraging. The need to identify and then sort out the relative importance of different drivers is a common feature of efforts to understand the behavior of complex and dynamic systems (Kasperson, Kasperson, and Turner 1995).

Case Studies

I have selected five cases for in-depth analysis of emergent patterns in international environmental governance: ozone, Antarctica, climate, whaling, and fur seals. All these cases feature familiar environmental and resource regimes operating at the international level. One reason for selecting them stems precisely from the fact that they are familiar. Because readers are apt to possess a basic understanding of these cases at the outset, there is no need to develop protracted accounts of the origins and history of individual arrangements before turning to the analysis of patterns of institutional change.

Most important, I have chosen these cases because they provide rich examples of the analytically distinct patterns of change identified and described in this chapter. Individual regimes can and often do feature changes that exemplify one pattern during an initial stage but shift to another pattern during later stages. There may be cases in which changes do not conform well to any of the patterns I have identified. But my goal in this analysis is to probe factors that give rise to a number of prominent patterns of institutional change rather than to assign a large number of cases definitively to individual categories on the basis of a simple taxonomy. The cases vary substantially on the dependent variable treated as emergent patterns of change. But they all involve regimes that have existed for some time (or, in one case, a regime that lasted for a long time before collapsing), a desirable characteristic given the focus of this analysis on emergent patterns in environmental governance rather than on a comparison between situations characterized by the presence or absence of regimes. There is also considerable variance among the cases with respect to both endogenous and exogenous conditions and the nature of the interaction between the two sets of conditions. Taken together, these circumstances provide reasonable assurance that selection bias will not afflict the analysis to follow. The results arising from the examination of a small number of cases cannot confirm specific hypotheses pertaining to the causes of patterns of institutional change. But the study of these cases can serve to enhance our understanding of the phenomenon of emergent patterns in environmental governance as well as lead to the formulation of new and important questions of interest to those conducting research on institutional change.

Stratospheric Ozone

Chapter 2 focuses on the evolution of the regime created to protect stratospheric ozone as an example of progressive development (Andersen and Sarma 2002; Parson 2003). The science relating to seasonal fluctuations in the thin layer of ozone in the stratosphere, especially in the high latitudes, evolved relatively slowly during the 1970s. But the growing influence of the theoretical arguments developed by atmospheric chemists like Paul Crutzen, Mario Molina, and Sherwood Roland, coupled with empirical observations of what we now know as the ozone hole, led to a striking succession of institutional developments starting in the mid-1980s. The 1985 Vienna Convention for the Protection of the Ozone Layer, a classic framework convention imposing few obligations on its members, was followed in 1987 by the Montreal Protocol on Substances That Deplete the Ozone Layer, an agreement committing signatories to reduce production and consumption of a number of CFCs and halons significantly and according to prescribed timetables. A number of additional developments adding strength to this regime have followed in a steady progression. The number of members has grown from 24 at the outset to 195, making this regime virtually universal. Amendments adopted in London (1990), Copenhagen (1992), Vienna (1995), Montreal (1997), Beijing (1999), and again in Montreal (2007) have accelerated phaseout schedules for ozone-depleting substances already covered by the regime and added several new families of chemicals to those subject to the provisions of the regime. In 1990, the parties to the regime created a mechanism known as the Montreal Protocol Multilateral Fund, which has played a significant role in inducing developing countries like China and India to join the regime and to comply with its provisions mandating the phaseout of a variety of chemicals. The regime has faced and continues to face significant challenges. But the evidence demonstrates that it has played a key role in reducing drastically the production and consumption of a sizable number of chemicals during its operation. It is one of the best candidates for the label of progressive development among the several hundred multilateral environmental agreements concluded in recent decades.

Antarctica

The Antarctic Treaty System, the subject of chapter 3, exemplifies the pattern I call punctuated equilibrium. Starting with the Antarctic Treaty itself, adopted in 1959 in the aftermath of the 1957–1958 International Geophysical Year and designed to demilitarize the south polar region and to alleviate tensions arising from jurisdictional claims in Antarctica, this regime has encountered a number of major challenges but responded in each case in such a way as to maintain and even enhance its effectiveness (Joyner 1998). The Antarctic Treaty is a stand-alone agreement whose twelve initial signatories, or Consultative Parties, shared a tradition of conducting scientific research in the area. Facing growing pressure during the 1970s and 1980s to admit new members to this "club" and even to integrate the regime into the UN System, the Consultative Parties opened the doors of the regime to a stream of new members and succeeded in preserving the status of the regime as a stand-alone arrangement. When interest rose in exploiting the living resources of the seas surrounding Antarctica and especially the large biomass of Antarctic krill (*Euphausia superba*), the treaty parties responded by negotiating the 1980 Convention on the Conservation of Antarctic Marine Living Resources, an agreement that is legally separate from the Antarctic Treaty but that is an integral component of the institutional complex that has become known as the Antarctic Treaty System (ATS). An even greater challenge arose during the 1980s in the form of growing tensions between those interested in searching for commercially significant deposits of mineral resources or hydrocarbons in Antarctica and those committed to setting the entire region aside as a protected natural area or, for all practical purposes, a wilderness area. The parties responded to this challenge initially by negotiating the 1988 Convention on the Regulation of Antarctic Mineral Resource Activities. But defections of several prominent members prevented this convention from entering into force. Key players in the system then moved promptly and vigorously to fill the resultant vacuum with the 1991 Environmental Protocol to the Antarctic Treaty, an agreement setting forth a substantial system of protective measures applicable to the whole of the south polar region. Thus, the regime for Antarctica has encountered a series of major challenges but has been able to respond to these challenges in ways that maintain or even strengthen the basic principles of the regime.

Climate

Chapter 4 turns to the regime designed to deal with climate change, an arrangement that differs in important respects from other regimes dealing with atmospheric issues (e.g., the protection of stratospheric ozone) and that exemplifies the pattern I call arrested development (Linden 2006). Both the significance of greenhouse gases in the Earth's atmosphere and the possibility of climate change resulting from a buildup of these gases have been understood for a long time. But concern about the dangers of climate change and especially the possibility of abrupt changes in the Earth's climate system is recent. As in the case of stratospheric ozone, efforts to address the problem of climate change began with the negotiation of a framework agreement, the 1992 UN Framework Convention on Climate Change (UNFCCC), followed five years later by the 1997 Kyoto Protocol to the convention. But here the similarities between international responses to these two atmospheric issues come to an end. The Kyoto Protocol did not enter into force until 2005. The United States, one of the two largest emitters of greenhouse gases, has not ratified the protocol and has not accepted any obligations to reduce emissions under the terms of this regime. Other large emitters including China, which has replaced the United States as the largest emitter of greenhouse gases, are treated as developing countries under the terms of the protocol and therefore exempted from accepting any obligation to reduce emissions. Countries with economies in transition—the former Soviet Union and its satellites—are allowed to take credit for reductions in emissions occurring as a result of economic collapse during the 1990s and even permitted to profit from the sale of what has become known accordingly as "hot air." The members of the European Union, acting as a bloc, may manage to meet their obligations under the Kyoto Protocol to reduce emissions by about 8 percent relative to 1990 levels during the course of the first commitment period (2008–2012). But few other UNFCCC Annex 1 countries—other than those with economies in transition—are expected to meet their reduction targets. The obligations of the Kyoto Protocol expire at the end of 2012, following the completion of the first commitment period. Negotiating the terms of a successor to the protocol has become the major item on the agenda of the Conference of the Parties to this regime. At the Bali meeting in 2007, the parties agreed to make a sustained effort to reach agreement on the terms of a successor to the Kyoto Protocol by the end of 2009. At this writing, consensus is lacking regarding a strategy that will attract effective participation on the part of large emitters, such as the United States and China, which together account for 35 to 40 percent of all greenhouse gas emissions. The description of the climate regime as a case of arrested development seems apt, at least for the moment.

Whales

The regime dealing with whales and whaling, the subject of chapter 5, exemplifies the pattern I call diversion (Friedheim 2001a). Major technological developments in the early twentieth century dramatically increased the capacity of human harvesters to kill great whales anywhere in the world. Initial regulatory responses during the 1930s made little headway and were cast aside during the war years. In 1946, the principal whaling nations entered into the International Convention for the Regulation of Whaling (ICRW), an agreement establishing a freestanding regime intended to promote sustainable harvesting of whales. Initially rather ineffectual, this regime began to gain traction during the 1970s by suspending the harvest of endangered species (e.g., the blue whale) and imposing meaningful quotas on the harvesting of other species. By the beginning of the 1980s, the regime was experiencing what developed into a state change. Several of the original signatories (e.g., the United Kingdom and the United States) had become nonwhaling nations, and in the meantime a number of other nonwhaling nations had acceded to the 1946 convention. In 1982, a coalition of members succeeded in mustering the three-fourths majority needed to pass a resolution within the International Whaling Commission to impose a moratorium on harvesting great whales starting with the 1985–1986 season and expected to last until the development of a Revised Management Procedure capable of ensuring the sustainability of future harvests. Leaders of the new majority sought to substitute the goal of preserving all whales for the goal of sustainable harvesting. But this did not lead to agreement on a new formulation of the regime's mission that could have produced institutional transformation. Although the balance of interests within this regime has shifted in recent years, the fact that the regime's decision rule requires a three-fourths majority to overturn the moratorium has made it impossible to reach agreement on some new way of characterizing the goals of the regime. The terms of the 1946 convention remain unchanged. But the regime for whales and whaling has emerged as a forum for ongoing battles between those desiring to recast the regime in a preservationist mold and those favoring a return to harvesting whales on a small-scale and highly regulated basis. The emergent pattern in this case features diversion followed by gridlock.

Fur Seals

Although many regimes survive on paper even after they are marginalized by changing circumstances or they become dead letters, others actually collapse and come to a more decisive end. Such a case, analyzed in chapter 6, is the four-party regime among Canada, Japan, Russia, and the United States governing the conservation and harvesting of northern fur seals (*Callorhinus ursinus*) in the Bering Sea (Gay 1987). This regime has a venerable history, starting with the adoption in 1911 of the North Pacific Sealing Convention, an international agreement widely credited with introducing a management system that led to a recovery of fur seal stocks and often regarded as a landmark in the evolution of efforts to protect or conserve wildlife at the international level (Lyster 1985). Although this international regime lapsed during World War II, the war deflected attention from the harvesting of seals and left the fur seal population intact. In 1957, the four parties formally reconstituted the regime for northern fur seals by reaching agreement on the Interim Convention for the Conservation of Northern Fur Seals, an arrangement requiring the parties to negotiate and accept periodic extensions of its terms after an initial period of twenty-two years. Commentators often pointed to the fur seal regime as a successful case of wildlife conservation at the international level. But major changes affecting the viability of this regime occurred during the 1970s and 1980s. Prominent among these were the creation of Exclusive Economic Zones under the terms of UNCLOS, the onset of large-scale but poorly understood biophysical changes in the Bering Sea ecosystem, the emergence of ecosystem-based management as an influential paradigm, and the rise of the preservationist movement among those interested in the fate of wildlife (NRC 1996; NRC 2003). In this setting, the United States opted not to ratify the 1984 protocol to the 1957 convention that would have extended the life of the fur seal regime for another four years. The regime that had lasted for decades and that had been effective during the early phases of its existence then collapsed almost overnight and sank without a trace.

Moving Ahead

The stage is now set for the empirical studies that follow. Each of these chapters explores a case that exemplifies one of the emergent patterns I have identified. They all have a common structure. Each study resembles a legal brief in the sense that it sets forth a fact pattern or a brief history of the case and then moves on to an analysis of the factors that explain the emergent pattern of change occurring in that case. Each study concludes with some observations regarding the road ahead for the regime, or, in the case of collapse, the prospects for the emergence of some new institutional arrangement in the foreseeable future. This structure leads to a certain amount of repetition in the sense that each chapter delves into key issues in both descriptive and analytic terms. Some readers may find this odd. But given the goal of this book, I regard this feature of the text as a strength rather than a weakness. Taken together, the case studies provide the raw material for the final chapter in which I revisit the endogenous-exogenous alignment thesis in general terms in the light of evidence drawn from the cases, discuss cutting-edge questions that will be of interest to those who think about emergent patterns of governance in environmental regimes, and comment on the policy relevance of the conclusions I reach.

2

Progressive Development: The Regime for Stratospheric Ozone

Overview: The Big Picture

Chlorofluorocarbons (CFCs), along with a number of other chemicals grouped under the rubric of ozone-depleting substances (ODSs), do not occur in nature. Developed in the late 1920s by an industrial chemist at General Motors, CFCs have a number of properties that make them attractive for a variety of industrial and commercial uses. They are nontoxic, nonflammable, and stable, all properties that are desirable particularly in comparison with chemicals previously used in a variety of applications (e.g., sulfur dioxide and ammonia). Before long, CFCs and related chemicals (e.g., halons) became popular as elements in many products including refrigerators, aerosol propellants, automobile air conditioners, cleaning solvents, blowing agents for insulation, and pesticides. Nothing was known at this stage about the role these chemicals would play as ozone-depleting substances. To many they seemed ideal—relatively easy to manufacture and considerably safer to use than their predecessors, such as ammonia in refrigerants.

All this changed in the postwar era in the form of a two-step process. First came a concern about threats to the integrity of the thin layer of ozone in the stratosphere. This concern was strong enough to play a role in stimulating opposition to supersonic transports (SSTs) during the 1960s and energizing the campaign to eliminate CFCs as aerosol propellants during the 1970s. But none of this involved a clear and widely accepted analysis of the cause of ozone depletion, much less an unambiguous argument pointing to anthropogenic forces in this realm. This second step emerged during the 1970s with the theoretical work of Crutzen, Molina, and Rowland that led in turn to the data collection and experiments that resulted in 1988 in a conclusive demonstration of the role of CFCs and other ozone-depleting substances as causal agents in this realm

(Parson 2003). Troubling as these findings were from the perspective of protecting the Earth's life support systems, tracing ozone depletion to human actions did have one positive implication. Human behavior is the one thing we can hope to influence as a matter of conscious choice, a fact that made it relevant to respond to the problem of ozone depletion by creating a regime designed to curb or redirect the human actions giving rise to this problem.

Factors that played an important role in pushing this issue to the forefront of the policy agenda include the predicted impacts of ozone depletion on human health and food security. Increased UVB radiation reaching the Earth's surface as a result of the thinning of stratospheric ozone is likely to damage human immune systems and to have negative impacts on major crops. The specter of sizable increases in the incidence of skin cancers, eye problems, and damaged crops gave this problem traction among members of the attentive public. Despite the fact that the impacts of increased UVB radiation are not uniform throughout the world, most people recognize that this problem belongs to the category of global environmental changes that are systemic in nature (Stern, Young, and Drukman 1992). The sources of the problem are global, and the potential impact on human health will be widespread, even though the most severe impacts will occur in the high latitudes of both hemispheres.

How should we think about ozone depletion as a problem calling for attention at the level of public policy? Three—by no means mutually exclusive—perspectives can help to illuminate the problem structure of ozone depletion as an international issue and provide insights that are helpful in thinking about effective ways to address the problem. One approach casts ozone depletion as a fairly straightforward externality problem. The release of CFCs and other ODSs into the Earth's atmosphere is an unintended and (initially, at least) unforeseen by-product of human actions designed to produce and consume safer refrigerators, better cleaning solvents, and so forth. To the extent that we take this view, the problem is familiar as a type of market failure calling for intervention on the part of government(s) to establish command-and-control regulations or create an incentive system like a cap-and-trade arrangement.

A second approach treats the ozone layer—and the atmosphere more generally—as a repository for the disposal of wastes arising from the production and consumption of goods valued by consumers in industrial societies. So long as they do not have to pay for this use of the atmosphere, producers and consumers are likely—implicitly, if not explicitly—to treat the atmosphere as a free factor of production and to use as much of it as possible in contrast to other factors (e.g., capital or labor) for which they do have to pay. The issue here centers on the need to recognize the atmosphere as a kind of public trust resource, the use of which should be managed or regulated by a responsible agency in much the same way that we deal with marine ecosystems that are affected significantly by human actions.

A third perspective focuses on strategies for fixing the problem. The stratospheric ozone layer is essentially a public good; an intact ozone layer is both nonexcludable and nonrival. If you take steps to improve the quality of the ozone layer, I will benefit from your efforts regardless of whether I make a contribution toward the supply of this public good. This perspective suggests that we should expect to see a good deal of free riding when it comes to addressing the problem of ozone depletion and that we will need to find ways to establish and administer an effective cost-sharing mechanism in this realm.

All three perspectives point to market failure as an essential feature of ozone depletion. Were this a small-scale local problem, it would make sense to explore possible solutions based on strategies designed to manage human actions in the absence of intervention on the part of government agencies. But given the global character of the problem, it would be naïve to count on the development of a spontaneous or self-generating collection of rules to solve this systemic problem. An obvious response to a situation of this kind is to negotiate a multilateral environmental agreement (MEA), while leaving primary responsibility for implementing the provisions of the regime to appropriate bodies located within individual member states. This is exactly what has happened in the case of the regime dealing with stratospheric ozone (Andersen and Sarma 2002). The 1985 Vienna Convention for the Protection of the Ozone Layer, the 1987 Montreal Protocol on Substances That Deplete the Ozone Layer, and major amendments to the protocol are all international agreements ratified by member states and accepted as legally binding international commitments. Primary responsibility for implementing these agreements on a day-to-day basis lies with designated domestic authorities. In the case of the United States, for instance, the provisions of the Montreal Protocol are implemented under the terms of Title VI of the Clean Air Act Amendments of 1990, a statute that designates the Environmental Protection Agency as the lead agency within the U.S. federal government for this purpose (Bryner 1993).

How has this regime worked in practice? As a consequence of the long periods of residence of ODSs in the Earth's atmosphere, the annual ozone holes over the high latitudes of the two hemispheres have not disappeared. Still, dramatic reductions in the production and consumption of many ODSs—up to 95 percent in a number of cases—have certainly slowed the buildup of ODSs in the atmosphere and should begin to take effect in alleviating the problem in the coming years. The members have strengthened the regime—accelerating phaseout schedules and adding new chemicals—on a steady basis as new information regarding both the nature and the consequences of the problem has become available. A notable recent development is the decision of the 2007 Meeting of the Parties (MOP) to the Montreal Protocol to accelerate the phaseout of hydrochlorofluorcarbons (HCFCs), which are greenhouse gases as wells as ODSs. This action will have the effect of limiting the generation of HFC-23, a by-product of the production of HCFCs and a powerful greenhouse gas. The regime is not without actual or potential problems. There is a black market in ODSs, and there are "banks" of ODSs in refrigerators and air conditioners manufactured before the regime took affect. What we can say at this point is that the regime has made a big difference in terms of outcomes or actor behavior; there is a good chance that it will prove effective in terms of problem solving over the longer run (Kaniaru 2007).

What is striking and very much in evidence in this case is the developmental path that the regime has followed. The 1987 Montreal Protocol followed the Vienna Convention in short order and before the emergence of full consensus regarding the biophysical mechanisms involved in the depletion of stratospheric ozone. The years following the agreement on the terms of the Montreal Protocol witnessed steady progress in taking vigorous steps to accelerate the phaseout of CFCs, adding new chemicals to the list of controlled substances under the terms of the protocol, and persuading most members of international society to become signatories to the agreements on which the regime rests. No dramatic setbacks or watershed changes have occurred during the life of this regime (Breitmeier, Young, and Zürn 2006). Few would deny that this case exemplifies the pattern of change I have labeled progressive development.

Facts: A Brief History of the Ozone Regime

A simple timeline (table 2.1) provides a good point of departure for understanding the ozone story. What stands out in this story is a pattern marked by some difficulty in getting a substantive regime up and running in the first place, followed by vigorous and effective efforts to strengthen

Table 2.1
Ozone Regime Timeline

1928	Invention of CFCs
1960s	Debates over supersonic transports
1974	Molina/Rowland model publicized
1977	Coordinating Committee on the Ozone Layer (CCOL) formed
1978	United States bans CFCs as aerosol propellants
1982	UNEP working group formed to prepare a convention
1985	Vienna Convention signed by 21 countries plus the EU
1985	Antarctic ozone hole announced
1986	International Ozone Trends Panel created
1987	Montreal Protocol signed by 24 countries plus the EU
1988	Ozone Trends Panel reports
1988	Vienna Convention enters into force
1989	Montreal Protocol enters into force
1990	MOP 2 accelerates phaseout schedules and creates MLF
1991	Ozone regime secretariat fully operational
1992	UN Conference on Environment and Development
1992	MOP 4 restricts HCFCs and methyl bromide
1995	MOP 7 tightens restrictions
1997	MOP 9—10th anniversary of Montreal Protocol
1999	MOP 11—phaseout schedules accelerated
2005	MOP 17—further cuts in uses of methyl bromide
2007	MOP 19—phaseout schedule for HCFCs accelerated
2008	Both the Vienna Convention and the Montreal Protocol have 195 signatories

the substantive provisions of the regime in response to new scientific evidence, a reliance on domestic agencies in the individual member states to implement the terms of the Montreal Protocol, and, most recently, the emergence of several new and somewhat different issues that will require serious consideration on the part of the MOP of the protocol in the near future.

As the table indicates, the process leading to the rise of the issue of ozone depletion to a high enough place on the international policy agenda to merit serious attention was not a simple one. The producers of CFCs and other ODSs—mostly large and influential multinational corporations—argued that it would be difficult, costly, and possibly harmful to public health to devise substitutes for these chemicals. The science regarding the consequences and especially the causes of ozone depletion was far from definitive. Conservative governments—such as the Reagan administration in the United States and the Thatcher government in the United Kingdom—came into power in key countries at critical moments. This did not prevent the emergence of a lively debate about the issue in many quarters and even some notable policy initiatives, such as the growing opposition to SSTs and the imposition of a ban on aerosol propellants using CFCs in the United States and a number of other countries. Yet these were somewhat desultory actions; they did not add up to the establishment of a comprehensive international regime with a mandate to take action to address the issue of ozone depletion.

Familiar accounts of the issue-attention cycle would lead us to expect that interest in ozone depletion would peak and then begin to fall off, despite the lack of a satisfactory resolution of the issue in policy terms (Downs 1972). But this did not happen. During the late 1970s and early 1980s, opinion in the scientific community began to converge around a well-defined model of the role of CFCs in the depletion of stratospheric ozone, first articulated explicitly in 1974 by Molina and Rowland. The fact that the issue raised serious questions pertaining to human health made it a matter of continuing public concern. The conservative governments in power in the United States and the United Kingdom may well have come to think of this issue as one allowing them to appear to be concerned with environmental issues without incurring major costs. And the publicity starting in the mid-1980s drawing attention to the annual occurrence of a large hole in the ozone layer over Antarctica provided a focusing event and trigger for taking meaningful steps to address the issue of ozone depletion (Kingdon 1995). Whatever the explanation, the United States took the lead in international negotiations aimed at launching a regime dedicated to coming to terms with the problem of ozone depletion (Benedick 1991/1998). The result was a sequence of events leading to agreement on the terms of the 1985 Vienna Convention and, in the process, touching off a remarkable round of institution building during the following years.

The Vienna Convention, a rather bland framework agreement, was followed in short order by the Montreal Protocol, in which the parties accepted quantified commitments to reduce the production and consumption of certain chemicals on an agreed-upon schedule. A series of amendments featuring the acceleration of phaseout schedules, the addition of new chemicals to the list of controlled substances, and the creation of a Multilateral Fund (MLF) followed in the succeeding years. By the turn of the century, the ozone regime had developed into one of the most mature members of the entire set of MEAs.

With respect to freezing and then phasing out the production and consumption of ODSs, the numbers tell a striking story (Parson 2003: 240–41). Advanced industrial members agreed under the terms of the Montreal Protocol to a 50 percent reduction in five major CFCs by 1998, together with a freeze on three major halons by 1992. At the London MOP in 1990, the parties reached agreement on an accelerated schedule for phasing out fifteen CFCs and three halons by 2000. They also added carbon tetrachloride and methyl chloroform to the list of controlled substances and took initial steps toward the regulation of HCFCs. In Copenhagen in 1992, members agreed to another acceleration with CFCs, halons, carbon tetrachloride, and methyl chloroform all slated for phaseout by 1996. They also agreed to freeze forty HCFCs by 1995, to freeze methyl bromide at 1995 levels, and to phase out thirty-four hydrobromofluorocarbons (HBFCs) by 1996. In Vienna in 1995, the parties agreed on further adjustments and reached consensus on a number of targets for reductions by developing (Article 5) countries, including a phaseout of CFCs and halons by 2010 and of methyl chloroform by 2015. At the MOP in Montreal in 1997, the parties decided to phase out methyl bromide by 2010, with Article 5 countries to follow by 2015. The Beijing meeting in 1999 brought additional developments, notably a freeze on production of HCFCs by 2004, with Article 5 countries to follow in 2016. The 2007 MOP produced agreement on an accelerated schedule for phasing out HCFCs.

In addition to this evidence of progressive development regarding phaseout schedules, other measures reinforce this picture of steady enhancement in the strength of the ozone regime. The parties took a number of steps to implement the provisions of the Montreal Protocol, featuring restrictions on trade with nonparties. In its initial version, the protocol calls for a ban on imports of ODSs from nonparties by 1990 and on products containing ODSs by 1993. The 1990 meeting resulted in a ban on ODS exports from all parties, an agreement extended to HBFCs (but not HCFCs) in Copenhagen in 1992. These trade restrictions became progressively tighter as the 1999 MOP extended the ban on imports and exports to include HCFCs and BCM (bromochloromethane). Overall, observers credit this regime with having brought about phaseouts of "95 percent of ozone-depleting substances in developed countries and 50–75 percent of ODS in developing countries" (Kaniaru et al. 2007: 1)

Another important element in the story of the ozone regime centers on issues of membership, including efforts to add parties to the regime and moves aimed at tightening the restrictions of the regime addressed to developing countries. Twenty-four countries signed the Montreal Protocol in September 1987. While this group included many of the members of the Organization for Economic Cooperation and Development (OECD), it did not include major developing countries like China and India. To address this problem, the parties agreed at the London meeting in 1990 to establish a Multilateral Fund with the goal of providing assistance to Article 5 countries willing to join the regime and accept commitments to phase out ODSs on a timetable acceptable to the developing countries. Starting with a pledge of $240 million, the MLF has received regular replenishments. As of 2009, contributions to the fund have totaled over $2.4 billion. This strategy worked. Today, 195 countries are signatories to both the Vienna Convention and the Montreal Protocol, making the regime virtually universal. It has taken time for countries to ratify more recent amendments. But this does not detract from the fact that this regime has become a universal governance system.

A ten-year grace period for developing countries and, more specifically, those whose per capita consumption of controlled substances was less than 0.3 kilograms per year also figured prominently in this developmental process. An early example of what has since become known as the principle of common but differentiated responsibilities, this formula, combined with the creation of the MLF, has played a role in the growth of the ozone regime to near-universal proportions. In becoming members of the regime, the Article 5 countries accepted some important provisions including the acceleration of phaseout schedules and the addition of new chemicals to the list of controlled substances, despite their special status under Article 5. This case testifies to the wisdom of differentiating among categories of members and adjusting the provisions of a regime to take into account differences in the circumstances of countries falling into separate categories.

Lest we lose perspective in thinking about the success of the ozone regime, several cautionary notes are in order. It is important to bear in mind the distinction between outcomes and impacts in this account. Because of their long periods of residency in the atmosphere, many ODSs will continue to affect stratospheric ozone for a long time to come. There is no guarantee that the phaseout of production and consumption of these controlled substances will lead in the final analysis to a simple restoration of the status quo ante regarding the stratosphere. A number of new issues pertaining to matters like illegal trade and CFC banks have come into focus. I have more to say about these emerging issues later in this chapter. For now, I mention them to ensure that we do not lose track of residual and emerging issues in focusing on the pattern of progressive development that is a hallmark of the case of ozone depletion.

Analysis: Sources of Progressive Development

How can we account for the progressive development of the ozone regime? To what extent can we extract lessons from experience with this regime that are relevant to other cases? While many specific factors played identifiable roles, the basic picture regarding institutional development in this case seems clear. A cluster of forces, including factors endogenous to the regime as well as exogenous factors, played important roles as drivers of institutional development. The confluence of these forces and, more specifically, the interplay between the internal and external forces emerge as the key to success in the development of the ozone regime from the embryonic version, articulated in the Vienna Convention, and the initial substantive version, encapsulated in the Montreal Protocol, to the more muscular arrangement that evolved during the 1990s, via the adoption of a series of amendments tightening restrictions on substances already on the controlled list and adding new chemicals to the list of controlled substances, as well as via the introduction of major new regime components like the Multilateral Fund.

Endogenous Factors

As chapter 1 indicates, endogenous factors may have consequences that are either developmental or degenerative in character. The same institutional feature may prove developmental under some conditions but degenerative under others. Voting rules offer a striking example. Rules that make it relatively easy to adjust a regime can contribute to progressive development in a setting conducive to progress. But the same rules can facilitate the introduction of degenerative changes in other settings. In the case of the ozone regime, these endogenous factors taken together fueled a developmental process that drove the regime forward in a distinctly progressive manner.

The shortest article—Article 18—in the Vienna Convention states in full that "[n]o reservations may be made to this convention," a provision that is carried over verbatim in Article 18 of the Montreal Protocol. This seemingly innocuous provision is actually a source of considerable strength in the case at hand. It is quite common for parties feeling ambivalent but not wanting to assume the role of spoiler to ratify an agreement, while simultaneously attaching various reservations to the act of ratification. Actions of this sort can and sometimes do compromise the problem-solving capacity of regimes at the outset. The inclusion of Article 18 of the convention and the protocol can be understood as a gamble that paid off with regard to the integrity and strength of the ozone regime. The prohibition on reservations has not become a stumbling block with regard to expanding the regime's membership. The ozone regime has achieved virtually universal participation among the members of international society, despite the Article 18 prohibition on reservations. As a result, the regime was able to get under way from a position of strength instead of finding itself crippled by reservations on the part of important member states and compelled to invest time and energy in finding ways to work around more or less significant limitations.

The ozone regime also features a progressive decision rule that is relevant to efforts to develop or strengthen the arrangement over time. Article 9(3) of the Vienna Convention states, "The Parties shall make every effort to reach agreement on any proposed amendment to this Convention by consensus. If all efforts at consensus have been exhausted, and no agreement reached, the amendment shall as a last resort be adopted by a three-fourths majority vote of the Parties present and voting at the meeting, and shall be submitted by the Depository to all Parties for ratification, approval or acceptance." Interestingly, this language is not replicated in the text of the Montreal Protocol, where the issue of amendments—especially to the reduction and phaseout provisions of Article 2—is a prominent concern. But Article 9(4) of the 1985 Convention is explicit in saying that the provisions of Article 9(3) shall apply to amendments to "any protocol, except that a two-thirds majority of the Parties to the protocol present and voting at the meeting shall suffice for their adoption." How important is this clearly progressive language when we come to the actual practice of the ozone regime? In collaboration with several colleagues, I have shown that consensus building—not to be confused with the search for unanimity—is a common practice in the operation of international environmental regimes, whether or not the use of a consensus rule is required under the formal provisions of the agreement (Breitmeier, Young, and Zürn 2006: ch. 4). This appears to be the practice that has emerged in the case of the ozone regime. It would be a mistake, therefore, to attach too much significance to the language of Article 9(3) as a factor that can help account for the progressive development of this regime. Yet the evidence suggests that the three-fourths and two-thirds decision rules have made a difference on some occasions in disciplining individual parties opposed to measures aimed at strengthening the provisions of this regime. While the practice that has grown up around the regime emphasizes consensual decision making, the implicit threat of resorting to these decision rules is always present.

Another endogenous factor that has facilitated progressive development in the case of the ozone regime involves the procedures governing the entry into force of amendments and especially amendments to Article 2 of the protocol dealing with control measures. Referring to such amendments as annexes, Article 10 of the convention states that these annexes "shall form an integral part of this Convention" and any protocols to it. While this article does allow individual parties to opt out of specific annexes, it provides for annexes to enter into force without explicit ratification on the part of individual member states. Thus, Article 10(2)(c) states, "On the expiry of six months from the date of circulation [of the required notice] the annex shall become effective for all Parties to this Convention or to any protocol concerned which have not submitted a notification" to the depository regarding their inability to approve the annex. The significance of this somewhat formulaic provision has turned out to be great. The parties to the ozone regime have interpreted this rule to mean that decisions to accelerate phaseout schedules for chemicals already covered under Article 2 of the protocol can enter into force without formal ratification on the part of member states, whereas a decision to add a chemical or family of chemicals not previously covered under Article 2 does require ratification. To take a concrete example, this means that the decision to accelerate the phaseout of CFCs and halons taken at the 1990 MOP in London took effect without going through a lengthy ratification process. But the decisions taken at the same meeting to add carbon tetrachloride and methyl chloroform to the list of controlled substances did call for ratification on the part of member states. This liberal provision regarding the acceleration of phaseout schedules has proven to be an important factor facilitating progressive development in the case of the ozone regime. Once chemicals are added to the list of controlled substances it is comparatively easy to respond to new scientific evidence regarding the role of these substances in causing a thinning of stratospheric ozone by accelerating phaseout schedules or making arrangements to apply such schedules to developing countries.

Another endogenous factor that has played a role in the progressive development of the ozone regime centers on the provisions governing the control of trade with nonparties set forth in Article 4 of the Montreal Protocol. This article deals both with imports of controlled substances from nonparties and with exports of such substances from member states to nonparties. Most imports were prohibited starting in 1990, with a ban on exports beginning in 1993. Although the actual provisions on imports and exports are somewhat complex, the members of the regime have ratcheted up restrictions on imports from and exports to nonparties steadily during the life of the regime. The provisions of Article 4 pertaining to nonmembers have become less and less relevant as membership in the ozone regime has approached universality. But this is properly interpreted as a measure of success in thinking about the progressive development of the regime. The regime's provisions relating to trade restrictions have played a role in making it unprofitable for nonparties to seek to reap benefits by manufacturing ODSs and selling them to member states or to provide for their own needs by importing controlled substances from regime members. This does not prevent nonparties from producing ODSs domestically and consuming them to meet their own needs. For most small countries, however, this has not been an option, since the production of CFCs, halons, and so forth requires a good deal of technological sophistication. This leaves the larger developing countries, such as China and India, as a focus of concern in the implementation and development of the ozone regime.

This is where the MLF, created as a new element of the regime at the 1990 MOP, comes into focus as an important endogenous factor. The MLF emerged as a mechanism for encouraging accession to the regime on the part of the larger Article 5 countries at a price acceptable to advanced industrial countries like the United States and the members of the European Union. A particularly interesting feature of this fund is the creation of a management structure quite distinct from the main ozone secretariat and highlighting the role of a fourteen-member executive committee in which half the members at any given time are representatives from the Article 5 countries. The MLF operates as a stand-alone arrangement, supplemented by Global Environment Facility (GEF) projects that are targeted specifically at countries with economies in transition. The funds allocated initially for the operation of the MLF—$240 million—have been replenished a number of times. Yet the overall cost of this mechanism for extending the scope of the ozone regime has been relatively modest—a total of about $2.4 billion from the inception of the fund to the present. It is not easy to make detailed cost-benefit calculations regarding the effects of resources made available through the MLF on the choices of key developing countries. But it seems clear that this mechanism has influenced the thinking of leaders in countries like China and India. This user-friendly funding mechanism has proven influential in allowing important developing countries to move directly to the consumption of substitutes rather than pursuing a strategy of moving first to the use of various controlled substances and only then making a transition to permissible substitutes.

Coupled with the restrictions on trade with nonparties and the creation of the MLF, what we now think of as the principle of common but differentiated responsibilities emerged as an endogenous factor promoting progressive development in the ozone regime. Article 5 of the Montreal Protocol, dealing with the special situation of developing countries, specifies in Section 1 that "Any Party that is a developing country and whose annual calculated level of consumption of the controlled substances in Annex A is less than 0.3 kilograms per capita on the date of the entry into force of the Protocol for it, or any time thereafter until 1 January 1999 shall, in order to meet its basic domestic needs, be entitled to delay for ten years its compliance with the control measures" adopted under Article 2 of the protocol. This is the provision that has produced regular references to the role of developing countries in accounts of the performance of the ozone regime. Article 5 goes on in a series of paragraphs to spell out a set of relatively detailed provisions governing the administration of the grace period for developing countries, including review procedures that call on the MOP to assess—not later than 1995—the situation of the developing countries and to direct particular attention to matters involving "financial co-operation and transfer of technology." The grace period for Article 5 countries is by no means an invitation to these countries to engage in a production and consumption derby with regard to ODSs. Reliance on the principle of common but differentiated responsibilities has made a difference in this case. It allowed developing countries to become members of the regime without accepting onerous obligations at the outset, while setting in motion a process of progressive development that would lead over time to the extension of provisions governing the phaseout of controlled substances to the Article 5 countries.

Several additional endogenous factors played significant roles in the progressive development of the ozone regime. The first centers on the articulation of goals that are explicit and, for the most part, measurable in quantitative terms. Calling for a 50 percent cut in current levels of production and consumption of five major CFCs by 1998, one of the goals articulated in the original 1987 version of the Montreal Protocol, provided a yardstick allowing for rigorous assessment of progress on the part of those responsible for the implementation of the protocol's provisions. Accelerating the schedule to a complete phaseout of fifteen CFCs by 2000, a goal set at the 1990 London meeting, made things even more straightforward. Any evidence of the production and consumption of these chemicals—apart from highly restricted quantities deemed acceptable for essential uses—on the part of member states from 2000 onward would constitute clear evidence of noncompliance. Other provisions, like the restrictions on trade with nonparties, are somewhat harder to evaluate. As a later secton of this chapter indicates, there is a black market in ODSs that is inevitably hard to assess, much less to suppress. Still, the adoption of measurable and especially quantifiable goals is a big help, especially in situations where progressive development leads to a process of revising these goals from time to time. It is true that the goals are framed in terms of outcomes in the sense of behavioral changes in contrast to impacts in the sense of problem solving. This opens the prospect that the goals could be fulfilled without leading to a complete restoration of the stratospheric ozone layer. But adopting measurable goals is a significant step in the right direction.

A particularly important feature of the ozone regime is the system of assessment panels established under Article 6 of the protocol calling on the parties starting in 1990 to "assess the control measures provided for in Article 2 . . . on the basis of available scientific, environmental, technical and economic information." Scientific assessment had played a major role already in the creation of the ozone regime (Parson 2003). In 1977, the United Nations Environment Programme (UNEP) initiated a process leading to the establishment of a Coordinating Committee on the Ozone Layer (CCOL). But in the context of this analysis of progressive development following the inception of the regime, the emphasis falls on a set of assessment panels set up under the authority of Article 6 of the protocol and operating from 1990 onward as three distinct bodies labeled the Technology and Economic Assessment Panel (TEAP), the Science Assessment Panel (SAP), and the Environmental Effects Assessment Panel (EEAP). The reports of these panels have become prominent features of the operation of the ozone regime. Mandated to engage in assessment in contrast to original research, the panels have played an important role in highlighting and drawing to the attention of the parties new knowledge (e.g., information regarding the availability of substitutes for various uses of ODSs) that has made a difference in terms of compliance with the provisions of the regime and in terms of attracting participation on the part of additional countries (Social Learning Group 2001). Assessment has flourished within this regime and played an important role not only in tracking compliance with the regime's requirements but also, and equally important, in helping the parties to come up with strategies for developing substitutes for ODSs at an acceptable cost.

This is also the place to observe that this regime broke new ground with regard to what have become known as noncompliance procedures (Victor 1998). Article 8 of the protocol calls on the parties to "consider and approve procedures and institutional mechanisms for determining non-compliance with the provisions of this Protocol and for treatment of parties found to be in non-compliance." What has emerged under this provision is a system that exemplifies what we now think of as a management approach to compliance (Chayes and Chayes 1995). The essence of this approach lies in an assumption that those failing to comply with the regime's requirements are apt to be doing so as a consequence of misunderstanding or a lack of capacity rather than as a matter of willful violation intended to reap benefits from the regime while avoiding the burdens of membership. Those adopting this approach generally recommend taking steps to work with parties that are failing to comply and to help them find ways to adjust their behavior to bring it into line with the requirements of the regime. The inability of the Russian Federation and a number of countries with economies in transition to meet obligations under the terms of the regime following the collapse of the Soviet Union, for instance, was treated as a problem to be solved rather than as an infraction to be punished. Efforts on the part of the regime's compliance committee with assistance from the Global Environment Facility resulted in a cooperative process in which these countries were able to return to compliance over a period of years (Andersen and Sarma 2002: 278–98). This does not prove that willful violations are rare or unimportant with regard to various provisions of the ozone regime. But the development of the regime's noncompliance procedures and its generally cooperative approach to dealing with noncompliance on the part of individual members is widely perceived as facilitating the process of progressive development.

Note as well that the ozone regime relies on the domestic practices of member countries with regard to the implementation of commitments to reduce or phase out the use of listed chemicals. It would have been possible, for instance, to require members to implement the provision calling for a 50 percent reduction in five major CFCs by 1998 through the creation of cap-and-trade systems in which emitters are required to be in possession of allowances or permits to emit, and the number of allowances available would decline by a prescribed amount each year (Tietenberg 2002). But there is no record of those responsible for the creation of the ozone regime and for its progressive development over time having considered the inclusion of such an international mandate. Why was this the case? It may be that such options were simply not on the table at the international level at that stage, though the U.S. Clean Air Act Amendments of 1990 had mandated just such a system for cutting sulfur dioxide emissions at the domestic level. It is possible that the creators of the ozone regime and those responsible for its development over time thought in terms of total bans or prohibitions from the outset and regarded the use of quasimarkets as unnecessary in an arrangement designed to produce total bans in relatively short order. Equally likely is the proposition that while it is appropriate to establish targets and timetables in an intergovernmental agreement, decisions regarding methods of implementation are best left to the discretion of individual members.

Although the destruction of stratospheric ozone is clearly a planetary concern, the creators of the ozone regime did not endeavor to draw everyone into this arrangement at the outset. Twenty-four countries signed the Montreal Protocol in September 1987; Article 16 of the protocol specifies that this agreement will enter into force with ratifications from as few as eleven countries, providing they represent "at least two-thirds of 1986 estimated consumption of the controlled substances. . . ." The strategy pursued in this case involved bringing a fairly small number of major players—specifically, the United States and some of the leading European countries—on board at the outset and assuming that ways could and would be found to draw in others following the entry into force of the protocol scheduled for January 1, 1989. It seems reasonable to treat this as an example of what Schelling has described as a "k-group strategy," starting with a small group of key actors and building out from this base to encompass others (Schelling 1978). In this case, it worked. Membership in the regime is now for all practical purposes universal. The choice of this strategy seems to run counter to the normal concern about free riders (Olson 1965). Would it not make sense for many (especially smaller) countries to treat the restoration of the ozone layer as a public good and seek to enjoy the benefits without shouldering a share of the burden? In the case of ozone, a combination of factors, including the threat of trade restrictions, the inclusion of a grace period, and the creation of the MLF effectively mitigated the free-rider problem. Circumstances vary from one case to another, but the ozone case does suggest that the k-group strategy can work in dealing with large-scale environmental issues.

It should be apparent at this point that the need for leadership on the part of dedicated and creative individuals looms just as large in the realm of progressive development over time as it does with regard to the earlier stage of regime formation (Young 1991; Young and Osherenko 1993). It would be a mistake to assume that matters like the operation of the assessment panels, the acceleration of phaseout schedules, the tightening of trade restrictions, and the development of noncompliance procedures would have occurred in the absence of innovative thinking and sophisticated bargaining on the part of dedicated individuals. In the case of the ozone regime, the role that Mostafa Tolba played, especially in the early years when he was executive director of UNEP, stands out in this regard. But this should not lead us to overlook the contributions of a number of others. With regard to the endogenous forces involved in progressive development, it seems correct to say that the emergence of effective leadership has been necessary to success, though of course it is not sufficient to produce such results.

Exogenous Factors

It would be wrong to infer from the argument I set forth in the preceding subsection that progressive development in the case of the ozone regime had a dynamic of its own that would have made it flourish regardless of what was taking place in the broader biophysical and socioeconomic settings. Starting with a range of factors affecting the scope and scale of the problem itself, we can identify a number of forces external to the provisions of the regime per se that made a difference in terms of promoting or hindering progressive development in this case. Among these are considerations relating to problem structure, cognitive factors, social considerations, economic forces, and political conditions.

Most environmental and resource regimes are created to solve particular problems. The biophysical and socioeconomic attributes of these problems are not parts of the institutional arrangements as such. But it is easy to see that important features of the problems under consideration can and often do have far-reaching consequences for the development of the institutions created to solve them. Several features of the problem facilitated progressive development in the case of the ozone regime.

Addressing ozone depletion is comparatively tractable in economic terms. The production of CFCs and halons had risen sharply to become a multibillion-dollar industry by the 1980s. But only a handful of companies were involved in the production of these chemicals; DuPont alone accounted for about 25 percent of the production. The sale of CFCs and halons was a small component of the activities of these firms in most cases; it amounted to only 2 to 3 percent of DuPont's annual volume of business. Although reliance on these chemicals for certain uses (e.g., refrigeration) had risen rapidly among advanced industrial countries, many countries had not yet become major consumers of CFCs and halons. Taken together, these factors helped to alleviate difficulties that arise when environmental policies impose heavy costs on small but easily identifiable and powerful interest groups (e.g., a handful of major firms in a single industry), while providing benefits to large but unorganized groups (e.g., members of the general public). None of the losers in this case would suffer irreparable harm. With regard to DuPont in particular, the company could proceed with considerable confidence on the assumption that it would emerge as a central player in the production and consumption of substitutes for ODSs.

Two additional features of the problem proved conducive to progressive development in this case. The potential harm arising from ozone depletion centered on impacts on human health (e.g., a rising incidence of skin cancers), a concern that always arouses interest among members of the general public. The revelation in 1985 of dramatic ozone holes occurring annually over the polar regions became a focusing event (Kingdon 1995); it had a shock effect that made ordinary people become aware of and alarmed by the prospect of ozone depletion. Skeptics could and did suggest that no harm would come to those who wore hats, sunglasses, and long-sleeved shirts. But such observations merely intensified the fear of increased UVB radiation in the minds of members of the general public. There has been some debate about the extent to which the revelation of the ozone hole influenced the negotiation of the specific provisions of the Montreal Protocol (Benedick 1991/1998). But fears about the health effects of increased UVB radiation played a significant role in generating public support for strengthening the provisions of the ozone regime in the period following its inception. The fact that the development of substitutes for most uses of ODS proved less difficult and less costly than many corporate players had predicted made it comparatively easy to respond to public fears about the potential impacts of ozone depletion on human health. In some cases (e.g., the development of cleaning solvents), equally effective and cheaper substitutes emerged from the need to come to terms with the phasing out of CFCs and halons. Overall, not only did the problem of ozone depletion generate substantial pressure to respond vigorously but also the character of the problem made it tractable to solve once the major players in the system acknowledged the need to do so.

The emergence of ozone depletion as a policy issue coincided roughly with the development of the analytic paradigm known as Earth system science. The effect of this was to provide a framework, or what some would call a new discourse, that made it possible to think of ozone depletion as a systemic problem at the planetary level requiring a vigorous response on a global scale (Litfin 1994). Science is no different from any other form of analysis in the sense that prevailing paradigms or discourses determine both the framing of issues for consideration and the allocation of time and energy among issues on the current agenda. It would have been difficult to focus on ozone depletion in a holistic fashion in the absence of the conception of the Earth as a complex and dynamic system that has become the hallmark of Earth system science (Steffen et al. 2004). In the absence of this systemic perspective, ozone depletion might well have been regarded as a regional phenomenon affecting sparsely settled areas (i.e., the polar regions) and having little impact on the politics of science (Greenberg 1999). As it turned out, the case of stratospheric ozone depletion emerged as a kind of poster child for the rapid growth of scientific interest in what we now routinely think of as global environmental change; an epistemic community concerned with this problem emerged and played a significant role in heightening the issue's salience as well as its legitimacy as a matter of public concern (Haas 1992). This not only brought the full weight of the scientific community to bear on the presentation of ozone depletion as an issue requiring serious attention on the part of policy makers but also made this case a test bed for emerging practices (e.g., scientific assessments) that have since become high profile activities in a variety of areas (e.g., climate change through the work of the Intergovernmental Panel on Climate Change) (Mitchell et al. 2006). The effort to deal with ozone depletion and the development of the science of global environmental change proceeded in tandem, a conjunction that had an impact on the amount of attention directed to the ozone problem in policy circles.

The fact that the anticipated impacts of ozone depletion centered on matters of human health served to enhance willingness on the part of policy makers and concerned citizens alike to make a concerted effort to identify the cause(s) of the problem and to take steps to do something about it. With all due respect to the importance of mainstream environmental problems that are seen as threatening biological diversity or imposing economic costs on those who become victims of externalities, issues perceived as likely to prove costly in terms of human health command unusual attention in policy circles. The fact that the first victims of ozone depletion would be residents of wealthy countries located in the antipodes (e.g., Australia, New Zealand, the Scandinavian countries, Canada, and the northernmost part of the United States) increased the pressure on policy makers to move the issue of ozone depletion to a high enough position on the policy agenda to merit explicit attention. Some countries (e.g., Australia) even added a UV index to daily weather forecasts that traditionally focus mainly on temperatures and the likelihood of precipitation. The problem of ozone depletion aroused a public desire to see policy makers do something significant to address the problem. The fact that the Reagan administration in the United States and the Thatcher government in the United Kingdom supported the adoption of the Montreal Protocol and took an interest in seeing it implemented in an effective manner is clear testimony to the social significance of the issue of ozone depletion.

The problem of ozone depletion has several economic characteristics in addition to those mentioned above that proved conducive to progressive development of the ozone regime. Ozone depletion shares with a number of other problems (e.g., acid rain) a feature in which the actual costs of taking effective action turn out to be substantially lower than the costs projected by key players in the debates prior to and leading up to regime formation. The sources of this phenomenon are not entirely clear. It may be attributable to path dependence in the behavior of companies that makes it hard for them to break away from existing production systems, even when there is the prospect of making profits by developing and marketing new products. It may be that corporate players believe that the use of new technologies will favor some companies over others, regardless of the overall benefits to society. Or necessity (in the form of legally binding commitments) may be the mother of invention in the sense that research and development efforts swing into high gear only after policy makers decide to impose restrictions affecting current products. A number of factors acting together probably account for this behavior. In the case of ozone depletion, the fact that leading economic players like DuPont were able to develop substitutes for ODSs that are not ruinously expensive not only made the ozone regime seem credible but it also cleared the way for an early acceleration of phaseout schedules for a raft of specific CFCs and halons.

The fact that the problem of ozone depletion did not lend itself to the use of standard cost-benefit calculations may also be significant. It is hard to quantify the incidence of health problems like skin cancers that may manifest themselves decades hence and that involve a number of causal forces. There is no simple metric for calculating the costs of these health effects both to the individuals affected and to society as a whole, and it is anyone's guess what an appropriate discount rate is in a case of this sort. Yet the specter of serious health effects and the existence of an aroused public provided policy makers with incentives to do something about the depletion of stratospheric ozone. The logical next step in such cases is to redirect attention from messy cost-benefit calculations to measures of cost effectiveness. The result is a policy dynamic that focuses on achieving a social goal as inexpensively as possible rather than on trying to decide whether achieving the goal is worth the cost of doing so. This leads to a search for new and less costly ways of pursuing well-defined goals selected as matters of public policy, a process that encourages innovation regarding ways to tackle a problem in contrast to recurrent and crippling debates about whether to pursue the relevant goal. Economic considerations, in a setting of this sort, become an engine for progressive development of institutional arrangements, like the ozone regime, in contrast to a drag on policy in the form of a way of thinking that leads to recurrent doubts about the value to society of addressing a problem in a manner that incurs substantial costs.

The broader political setting, too, was conducive to progressive development in the case of ozone depletion. In contrast to its performance in other issue areas, the United States exhibited leadership in the formation and development of the ozone regime. The development of the regime became a matter of negotiations between the United States and the European Union, both of which expected to be affected by the seasonal thinning of stratospheric ozone and were in a position to take effective steps to address this problem, so long as the price was reasonable. As noted, this case proved appealing even to conservative governments that saw in it an opportunity to demonstrate their concern for the environment without incurring political commitments that would detract from their policy agendas. The effort to curb ODSs did not generate an extensive debate about the relative merits of alternative policy instruments like the current discussion of the pros and cons of cap-and-trade systems versus carbon taxes or various regulatory measures as procedures for coming to terms with climate change. The regime emerging from the Vienna Convention and the Montreal Protocol took shape during a period in which those responsible for institutional design, especially at the international level, were content to leave such matters to the preferences of individual member states. It is impossible to tell at this point whether the creation of a cap-and-trade arrangement covering CFCs and halons would have made sense as a mechanism for phasing out the use of ODSs in an efficient manner. But it seems probable that the absence of a protracted debate about policy instruments in this case actually made progressive development easier than it would have been otherwise.

Endogenous-Exogenous Alignment

Some analysts direct attention to problem structure in their efforts to think systematically about institutional change. They argue that it is possible to rank problems on a scale that runs from highly benign to highly malign and that success in problem solving will be determined in large measure by the locus of a specific problem on this scale (Rittberger 1990; Young 1999a; Miles et al. 2002). Those who adopt this line of analysis are apt to argue that the problem of ozone depletion was relatively benign. The uses of CFCs and halons were limited to specific economic sectors, there was an aroused public calling for serious steps to address the problem, and key actors found it in their interests to take action regarding this problem. It follows, on this account, that we should not be surprised by the occurrence of progressive development regarding institutional responses to ozone depletion.

How persuasive is this line of reasoning? Not only do we lack an operational measure of problem structure but there are also reasons to question this interpretation of progressive development in the case of ozone depletion. A relatively high degree of uncertainty surrounded this issue. A concentrated but powerful industry resisted action, whereas those calling for action constituted a large but poorly organized group. At the time, the idea that global environmental changes call for global responses was just coming into focus. Some problems are undoubtedly harder to solve than others. As I argue in chapter 4, the problem of climate change has features that make it peculiarly difficult to address effectively. But it would be a mistake to pronounce that the problem of ozone depletion was benign and therefore easy to solve using familiar methods.

The essential message emerging from a careful assessment of progressive development in the case of ozone depletion is a sense of the importance of the fit between the problem to be solved and the capacity of the institutional arrangements created to address it (Miles et al. 2002; Young 2002; Young, King, and Schroeder 2008). To be successful, those responsible for creating and implementing environmental regimes need to concentrate on developing arrangements that are well suited to the character of the problem at hand and to salient features of the broader biophysical and socioeconomic settings. Even a problem that seems comparatively easy to solve can prove resistant when the relevant regime is flawed in the sense that it is not tailored to the situation at hand. Conversely, problems that most observers would classify as hard to solve and therefore malign can give way to actions taken in conjunction with a regime carefully crafted to fit the case at hand. This suggests the importance of asking questions about complementarities between endogenous factors and exogenous factors in the search for explanations that can account for institutional development following the stage of regime formation.

This endogenous-exogenous interface proved to be an important source of progressive development in the case of ozone depletion. A few concrete examples will serve to illustrate this phenomenon and to demonstrate its importance with regard to this case. The regime strikes a balance between articulating requirements that members are expected to meet and maintaining flexibility in the face of considerable uncertainty about important features of the problem. There is no denying the role of uncertainty in this case. At the time of the adoption of the Montreal Protocol in September 1987, the scientific community had not reached a definite consensus regarding the validity of the Molina/Rowland analysis of the role of CFCs and halons in destroying ozone in the stratosphere. Even after consensus emerged on this core issue in 1988, new insights regarding the effects of CFCs and halons and the role of additional chemicals (e.g., carbon tetrachloride and methyl chloroform) in this context arose in a steady stream. The genius of the regime in this regard lies in the provisions allowing for the acceleration of phaseout schedules without a requirement for ratification on the part of member states, while requiring ratification of decisions calling for the inclusion of additional chemicals. The role of this regime as a poster child for new thinking about Earth system science is worth noting in this context as well. Due to the regime's exemplary role in this realm, both the problem and the regime created to address it became a focus of sustained interest in the scientific community and in the policy community. This led to the establishment in 1977 of the Coordinating Committee on the Ozone Layer under the auspices of UNEP and to marked progress in the development of what we now understand as scientific assessment (Parson 2003). This resulted in turn in the creation under the terms of the Montreal Protocol of the Scientific Assessment Panel, which promptly began to gather data regarding the phenomenon of ozone depletion that allowed it to issue an initial report already in 1989 characterizing the commitments undertaken in the Montreal Protocol as inadequate to solve the problem of stratospheric ozone depletion.

Similar remarks are in order regarding the measures adopted to address the problem of ozone depletion. The potential for free-ridership in conjunction with efforts to solve the problem is evident. In the absence of side payments, it is easy to imagine that many countries and especially large Article 5 countries like China and India would have sought to enjoy the benefits arising from the initiatives of others in this realm, without accepting obligations themselves. The Multilateral Fund, added to the regime in 1990, effectively eliminated this problem by providing potential free riders with financial incentives sufficient to bring them into the regime. The cost of this solution to the free-rider problem was significant: the advanced industrialized countries have committed about $2.4 billion to the MLF since its inception. As a solution to the problem of building capacity among the Article 5 countries and persuading them to come on board as active participants in the campaign to protect stratospheric ozone, however, there can be no doubt that the development of the MLF—endowed with substantial funds and a decision-making procedure friendly to the developing countries—was a cost-effective measure that made an important contribution to progressive development.

The choice of policy instruments also made a difference in terms of progressive development. Casting this arrangement fundamentally as a prohibition regime and calling on members to phase out the production and consumption of a range of industrial chemicals entirely has helped to make the basic character of the arrangement clear. Individual members can make their own decisions about how to meet phaseout schedules adopted under the terms of the ozone regime. But it is difficult to see any role in this case for offsets of the sort that have become familiar in efforts to address the problem of climate change. A major part of the response to the problem of ozone depletion has centered on providing developing countries with opportunities to bypass the production and consumption of ODSs in contrast to phasing out existing uses of the relevant chemicals. This does not warrant drawing analogies between the case of ozone depletion and other cases involving the control of pollutants like acid rain or climate change. But one source of progressive development in the ozone case has been the emergence of a good match between the policy instruments selected and the economic and political circumstances of those subject to the regime's provisions.

Forecast: The Road Ahead

The case for declaring the regime to protect stratospheric ozone a success is strong. As figure 2.1 indicates, levels of ODSs in the Earth's atmosphere are well below what they would have been in the absence of this regime. The regime has developed in a progressive manner. Key events during the life of the regime include the acceleration of phaseout schedules for chemicals already covered, the addition of new chemicals to the roster of controlled substances, the establishment of the MLF, and the expansion of membership to near universal proportions. Still, it is important to bear in mind the distinction between outcomes and impacts in this connection. Reductions in the production of ODSs are relatively easy to document, and the operation of the MLF has been a success in persuading Article 5 countries to join the regime. But this does not mean that the problem has gone away. The seasonal thinning of stratospheric ozone is beginning only now to decline in a measurable fashion. It is a simple matter to explain this in terms of the long residency in the atmosphere of prior emissions. But this makes it important to withhold any final judgment on the performance of this regime until we see what happens with regard to the seasonal thinning of stratospheric ozone over the next two to three decades.

Beyond this lies the issue of emerging or foreseeable threats to the performance of the ozone regime. Granted, progressive development has been the hallmark of this regime so far. But what lies in store for the regime in the coming years? We cannot answer this question in any definitive way at this point. What we can do is ask questions about factors likely to influence the fate of the ozone regime over the next couple decades. Perhaps the best way to approach this matter is to look for emerging issues that have the potential to pose serious problems for the effectiveness of the regime. Three such issues stand out: the development

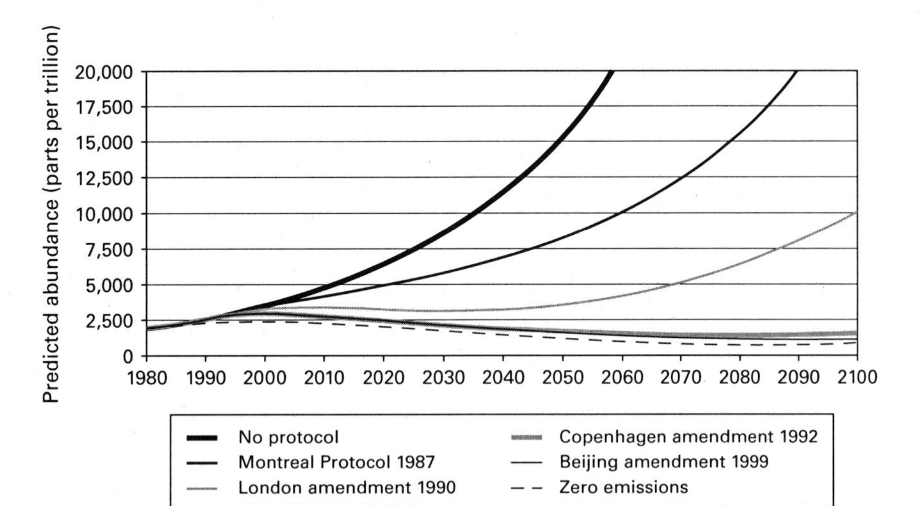

Figure 2.1
Effect of international agreements on the predicted abundance of ODS in the stratosphere 1980–2100. *Source*: GEO Data Portal, compiled from UNEP-Ozone Secretariat 2006, UNEP Geo-4 Report, October 2007, http://www.unep.org/geo/geo4/media/graphics/zoom/2.25.jpg.

of a black market in CFCs, the potential release of banked ODSs, and the interplay between this regime and others, particularly the climate regime.

There has been a black market in ODSs for some time. Although it is difficult to judge the scope of illegal trade in ODSs, it is not trivial. Kaniaru and colleagues (2007: 8) say that "[i]llegal trade currently is estimated to represent about 10–20 percent of all trade in ODSs, which in CFCs alone comprises 7,000–14,000 tons per year, with a value of U.S. $25–60 million." Is this a big number reflecting a serious threat to the effectiveness of the ozone regime? The answer to this question rests on two factors: the violation tolerance of the regime and the prospect of further development of the black market as restrictions on legal trade in ODSs become more stringent and take effect with regard to Article 5 countries. All regimes have to deal with problems of compliance, but they differ greatly in terms of their violation tolerance (Young 1999a: ch. 4). In the case of the ozone regime, violation tolerance is relatively high. Unlike the ban on nuclear weapons testing, which is not likely to survive even one major violation, the key concern with regard to the ozone regime is a matter of the total ODSs entering the Earth's atmosphere. So long as the total exhibits a steady decline, the regime is likely to be seen as successful. The 2007 agreement on the part of Article 5 countries to take initial steps to phase out HCFCs by 2015 with final phaseout by 2030 is an important turning point in this connection.

The idea of banked ODSs refers to residual chemicals in products like refrigerators manufactured prior to the phaseouts mandated under the Montreal Protocol as amended at various MOP meetings. As Kaniaru and colleagues (2007: 5) explain it: "Banks are defined as the chemicals contained in equipment and products or stored in tanks. Large amounts of CFCs and other ODS substitutes such as HCFCs and HFCs . . . currently exist in refrigerators, air conditioners, insulating foam, and chemical stockpiles, where they can leak. When equipment reaches the end of its useful life, the chemicals inside are usually released into the atmosphere." These banked ODSs could well prove to be a major problem in terms of the goal of the ozone regime to protect the stratospheric ozone layer from further thinning and to restore this layer to its preexisting state over a period of several decades. But this problem seems to be an excellent candidate for treatment through the process of progressive development that is a prominent feature of this regime. To do so may require a reprogramming and further replenishment of the MLF, especially to assist Article 5 countries to develop the capacity and the incentives needed to capture ODSs in old refrigerators and other outdated equipment before they begin to leak into the atmosphere on a sizable scale. But this does not seem to be an insuperable problem.

The issue of interplay between the ozone regime and other regimes is somewhat different in character. The most prominent case in point at this time involves interaction between the ozone regime and the climate regime. Under the terms of the Montreal Protocol as amended, the use of HCFCs as a substitute for CFCs has been permissible. Amendments to the protocol have required developed countries to freeze production by 2004. But even under the 2007 amendments, Article 5 countries have until 2015 to start reducing production of HCFCs. This has given rise to perverse incentives with regard to HFC-23, a powerful greenhouse gas that occurs as a natural by-product of HCFC-22. To take a concrete example, members of the EU's Emissions Trading Scheme have been able to purchase greenhouse gas offsets by paying for the destruction of HFC-23 under the provisions of the Kyoto Protocol's Clean Development Mechanism. This has created an incentive for certain countries to manufacture HCFC-22 at least in part to receive compensation for destroying the HFC-23 by-product rather than to fulfill a genuine need for ODS substitutes. Participants in the twentieth anniversary session of the ozone regime's MOP in Montreal in 2007 took a major step forward in agreeing on measures to close this loophole. Yet this situation exemplifies the sorts of difficulties that can and likely will occur with growing frequency as we move toward the creation of regimes to cope with a variety of large-scale and interrelated environmental problems.

It is possible already to foresee potential conflicts between the ozone regime and other arrangements dealing with health, food systems, and trade. Because the probable impacts of the depletion of stratospheric ozone include harm to human health and damage to food crops, it is relevant to ask whether the World Health Organization (WHO) or the UN Food and Agriculture Organization (FAO) should have a say in determining which ODSs to include on the list of controlled substances and what phaseout schedules are appropriate with regard to specific ODSs. Similarly, since the ozone regime relies on trade restrictions as a means of putting pressure on nonmembers, it is relevant to inquire about the role of the World Trade Organization (WTO) in decisions about the use of such measures in specific cases. We should not exaggerate the significance of these instances of institutional interplay. The ozone regime has accelerated phaseout schedules even in the absence of formal input from WHO, FAO, and WTO. Because membership in the regime is now virtually universal, the issue of trade with nonparties has largely evaporated. Still, the issue of institutional interplay is important in any examination of emergent patterns in specific institutional arrangements. Some cases of interplay are typically positive or even synergistic rather than conflictual (Oberthür and Gehring 2006). But the significance of institutional interplay as a factor likely to play a role in institutional dynamics is certain to grow in a setting featuring increases both in the numbers and the stringency of environmental regimes (Young, King, and Schroeder 2008).

Conclusion

So far, at least, the regime created to protect stratospheric ozone exemplifies the pattern of change I call progressive development. The formal components of the regime have evolved from a framework convention to a substantive protocol and on to a series of amendments tightening the restrictions on the production and consumption of ODSs. Although the ozone layer itself has not yet recovered from the damage caused by past emissions of ODSs, measures are now in place to halt the destruction of stratospheric ozone and to restore the ozone layer to something like its former state over a period of several decades. There are identifiable obstacles that may impede efforts to achieve this goal. The existence, at least for the near future, of a black market in ODSs and the presence of sizable quantities of ODSs banked in older or discarded air conditioners, refrigerators, and so forth seem particularly important in this connection. There may be additional challenges that have not yet become apparent as problems confronting the ozone regime. Large biophysical systems cannot be counted on to revert to the status quo ante once perturbations are eliminated. Still, none of this can or should detract from the remarkable development of this regime over a period of several decades. The regime has evolved from the embryonic version articulated in the 1985 Vienna Convention to a control mechanism capable of phasing out or banning a range of chemicals implicated in the thinning of the ozone layer.

What are the sources of this pattern of progressive development in the evolution of the ozone regime? The answer lies in the juxtaposition of factors that are endogenous and factors that are exogenous to the regime. The ability of the regime to accelerate phaseout schedules, add new chemicals to the list of controlled substances, provide incentives for nonparties to join the regime, deal constructively with cases of noncompliance, and assess performance systematically has contributed to the progressive development of this governance system. But exogenous factors have played significant roles as well. Although it is difficult to place environmental problems on a simple benign-malign scale and to forecast prospects for the success of specific regimes on this basis, some features of the problem (e.g., the fact that the production and sale of CFCs and halons was a small part of the business of a few major corporations and that these corporations found it relatively easy to develop substitutes for ODSs in most applications) facilitated efforts to move forward in the development of the ozone regime. Other exogenous factors played a role in this case as well. The growth of Earth system science as a new framework for thinking about large-scale environmental issues, the sensitivity of the public to issues presented as public health concerns, and the fact that conservative governments like the Reagan administration in the United States and the Thatcher government in the United Kingdom found reasons to promote the regime all contributed to the emergence of a pattern of progressive development in this case. Leadership on the part of individuals also played a role. We have been aware for some time of the role of individuals in the context of regime formation (Young and Osherenko 1993). With regard to ozone, researchers have documented leadership on the part of individuals like Mostafa Tolba, the head of UNEP; Winfried Lang, the chairman of the international negotiating committee; and Richard Benedick, the lead negotiator for the United States, in negotiating the terms of the Montreal Protocol. Leadership on the part of a number of people, especially with regard to the progressive steps of the 1990s, also made a difference. It would be wrong to look to the ozone case as a source of simple recipes for those desiring to nurture a pattern of progressive development in other issue areas. The interplay of an array of endogenous and exogenous factors is much too complex to allow for that. Still, this case does provide us with a clear illustration of the sorts of institutional dynamics that deserve the label progressive development.

3

Punctuated Equilibrium: The Antarctic Treaty System

Overview: The Big Picture

Antarctica, the Earth's most remote continent, has no permanent human residents. Undiscovered until well into the nineteenth century, the continent today has a human population limited to the occupants of a collection of widely scattered research stations that numbers some five to seven thousand during the austral summer but dwindles to less than a thousand during the winter months. As a result, some of the familiar sources of competition and conflict in human affairs, such as externalities arising from the activities of neighbors whose actions affect one another's welfare, are irrelevant in the Antarctic. Still, it would be a mistake to infer from this that there are no important and high-profile issues of governance arising in the south polar region. Among the relatively rich array of governance issues are those pertaining to the specification of criteria for consultative-party status in the Antarctic Treaty System (ATS), the assertion of jurisdictional claims on the part of so-called claimant nations, the maintenance of Antarctica as a demilitarized area, the need to protect the continent's wildlife and fragile ecosystems more generally, the regulation of fishing in the waters surrounding Antarctica, the treatment of plans to engage in exploration or prospecting for minerals, the administration of science in and around Antarctica, and the management of Antarctic tourism. Despite the region's remoteness, therefore, there is no shortage of issues that generate a demand for the creation and implementation of governance systems or for the periodic adjustment of existing arrangements to cope with changing circumstances (Rayfuse 2007).

All efforts to meet the demand for governance in Antarctica must come to terms with some anomalous circumstances that can and often do cause confusion among those interested in problems of governance. There is a widespread tendency to treat Antarctica as a part of the global commons in much the same way that we consider the high seas, the deep seabed, outer space, and the moon as commons belonging to what has become known as the common heritage of humankind (Joyner 1998). But what does this mean in practice? Unlike the high seas or outer space, most of Antarctica is subject to (sometimes overlapping) jurisdictional claims on the part of seven claimant states. Major nonclaimant states (e.g., the United States and Russia) also assert relatively well-developed and extensive use rights in the region that cut across the jurisdictional claims of the claimant states. The vision of Antarctica as a form of international common property where human activities are subject exclusively to rights, rules, and decision-making procedures devised by the members of international society is cloudy.

An alternative interpretation suggests that we should think of Antarctica as a large-scale common-pool resource in the sense of an area that is open to entry on the part of users whose activities are in some measure rival (Young 2007. Here, too, complications set in immediately. Antarctica is not open to entry on the whim of potential users. The logistical requirements for use exceed the capacity of many potential users; even more important is the fact that the ATS regulates uses of Antarctica and its resources on the part of public and private actors in a relatively strict manner. Nor is it apparent that all uses of Antarctica are rival in the ordinary sense of the term. So long as activity levels remain limited in scope and scale, a diverse set of users can conduct scientific research, engage in well-regulated tourism, and take steps to protect vast segments of the region as wilderness areas without imposing costs on one another or generating serious conflicts of interest arising from unintended side effects of any of these uses.

What is needed in the case of Antarctica is an approach to governance that eschews many of the simple conceptual categories we normally employ in thinking about such matters and that is uniquely matched to both the biophysical conditions and the human dimensions of this special region (Young 2002). The system that has arisen and evolved over the past fifty years, with the 1959 Antarctic Treaty as its core and known generally as the Antarctic Treaty System, fulfills this requirement in an exemplary fashion. It sidesteps potentially conflicting issues in an ingenious manner, takes advantage of the flexibility inherent in an institutional complex, and exhibits a striking ability to adapt to changing circumstances without jeopardizing or discarding its defining features.

The master stroke of the Antarctic Treaty centers on the introduction of an interlocking set of arrangements that neutralize jurisdictional issues, exempt the region from geopolitical maneuvering between or among the great powers, and take advantage of the growth of Antarctic science as a means of ensuring transparency in activities relevant to governance in the south polar region. An arrangement of this sort might prove ineffective or might even exacerbate tensions in regions characterized by high levels of human activity. But the device of freezing jurisdictional claims by stipulating that activities carried out under the terms of the Antarctic regime will neither enhance nor diminish existing claims has allowed the claimant states to stand their ground, while opening up the whole continent to activities carried out by scientists and other authorized personnel in the absence of the administrative or bureaucratic restrictions that plague many governance systems. The treaty's designation of Antarctica as a continent for peace, effectively demilitarizing the entire region, has limited sensitivities regarding the activities of claimant and nonclaimant states alike and, in the process, made it relatively easy for nationals of all treaty members to go about their business in Antarctica without engendering suspicions or provoking efforts to impose debilitating restrictions on their activities. The role of science as the principal human activity on the continent has helped to sustain these cooperative arrangements (Berkman 2002). Because science is widely regarded as an international activity that is largely free of political biases, scientists have been able to carry out their work in Antarctica wherever it takes them and, in the process, to lend substance to the principle that the region should be open to all on a nondiscriminatory basis and to enhance the transparency of the activities of the treaty parties in and around Antarctica.

The ATS is an institutional complex in the sense that it is composed of a number of discrete elements that are tightly linked but that do not constitute a single integrated constitutive system (Raustiala and Victor 2004). The Antarctic Treaty is the core element of this complex, a status that has been substantially strengthened by the addition of the 1991 Environmental Protocol with its explicit vision of the ultimate goal of ensuring that Antarctic ecosystems remain intact over the long haul. Other elements of the complex include formal international agreements like the 1972 Convention on the Conservation of Antarctic Seals (CCAS) and the 1980 Convention on the Conservation of Antarctic Marine Living Resources (CCAMLR), regulatory measures intended to implement the general provisions of the system (e.g., the 1964 Agreed Measures on the Conservation of Antarctic Fauna and Flora), and agreements among nongovernmental organizations (e.g., the code of conduct covering the activities of Antarctic tour operators). Day-to-day practice increasingly has had the effect of welding these component parts into the coherent whole we generally have in mind when speaking of the Antarctic Treaty System, and it is fair to say that the whole system now operates under a common commitment to ecosystem-based management (EBM). Yet the arrangement does introduce a measure of flexibility in handling a range of concrete issues arising in the region. It is possible to address matters relating to the harvesting of fish, for example, without turning them into issues requiring the mobilization of the entire apparatus of governance dealing with Antarctic issues.

A striking feature of the ATS is its ability to respond to major challenges—some may even regard them as crises or watershed events—in an effective manner and, in some instances, in a manner that actually strengthens the whole system. Prominent examples include intense criticism of the ATS as an exclusionary club during the 1960s and 1970s, the growth of interest in commercial harvesting of marine living resources in the south polar region during the 1970s and 1980s, and the conflict between those interested in prospecting for commercially valuable minerals in the region and those dedicated to preserving Antarctica as a wilderness area during the 1980s and 1990s. The ATS has responded to these challenges in a manner that demonstrates a high level of resilience. The admission of new consultative parties has not triggered any dramatic expansion of human activities in the region. Fishing has remained relatively limited in Antarctic waters, especially in comparison with the expectations of those who envisioned the development of immense harvests of krill (*Euphasia superba*) in the region. The low probability of discovering mineral deposits large enough to make them attractive to commercial operators despite the remoteness and the harsh conditions of the area played a role in the acceptance by all parties concerned of the demise of the 1988 Convention on the Regulation of Antarctic Mineral Resource Activities (CRAMRA) and the decision to replace it with the 1991 Environmental Protocol based on a strikingly different set of principles. The occurrence of this institutional volatility should not be allowed to diminish the record of the ATS as a highly successful governance system. This striking pattern of challenge and response is the hallmark of the system's behavior. It justifies the conclusion that this case fits the pattern I have labeled punctuated equilibrium (Repetto 2006).

Facts: A Brief History of the Antarctic Treaty System

Twelve states signed the Antarctic Treaty on December 1, 1959. Seven of the twelve (Argentina, Australia, Chile, France, New Zealand, Norway, and the United Kingdom) maintained jurisdictional claims in Antarctica and have become known as the claimant states. The other five (Belgium, Japan, South Africa, the Soviet Union, and the United States) maintained a right of access everywhere in Antarctica and are known as the nonclaimant states. The regime established under the terms of the 1959 treaty did not begin life as an environmental or resource regime. The essential bargain struck by those negotiating the 1959 Antarctic Treaty is embedded in the provisions of Article I specifying that "Antarctica shall be used for peaceful purposes only" together with those of Article IV freezing jurisdictional claims and stating, "No new claim, or enlargement of an existing claim, to territorial sovereignty shall be asserted while the present Treaty is in force." The treaty, which entered into force in 1961, thus served the interests of the superpowers (the United States and the then Soviet Union) in securing access to the whole continent while exempting the region from the competitiveness of the cold war. The treaty also protected the interests of the claimant states by recognizing their existing claims and in the cases of those states asserting overlapping claims—Argentina, Chile, and the United Kingdom—by sustaining their claims without engendering any pressure to resolve their differences.

A number of actors and most prominently the United States had pushed for the adoption of some sort of governance system applicable to the south polar region since the late 1940s without success (Peterson 1988). What made the difference in 1959 was the emergence of science as an important player in Antarctica, as manifested in the range of activities undertaken in the course of the 1957–1958 International Geophysical Year (IGY) (Belanger 2006). This is not to imply that science had much to contribute to the institutional bargaining leading to the adoption of the Antarctic Treaty. Rather, science became a useful tool for those seeking to address geopolitical concerns relating to Antarctica not only by providing an organized activity through which states could demonstrate their ongoing interest in the region but also by playing a noncontroversial role in operationalizing the provisions of Article VI of the treaty, dealing with access to Antarctic stations for purposes of observation, inspection, and monitoring. Because the treaty describes the criterion for consultative-party status in this arrangement as the conduct of "substantial scientific research activity" in Antarctica, science also emerged as a common denominator creating shared interests and playing a key role in implementing treaty provisions in a cooperative manner. The unusually strong links that have developed between the ATS and the Scientific Committee on Antarctic Research (SCAR)—a body operating under the auspices of the International Council of Science (ICSU)—testifies to this mutually beneficial relationship.

The Antarctic Treaty contains only one reference to environmental or resource issues.[1] This is the provision of Article IX(1)(f), authorizing the consultative parties to adopt measures "in furtherance of the principles and objectives of the Treaty" dealing with the "preservation and conservation of living resources in Antarctica." But as the timeline set forth in table 3.1 makes clear, this situation has changed dramatically over the life of the regime. It is not that demilitarization and the freeze on jurisdictional claims are no longer relevant. They provide the essential foundation for the growth of a dense cooperative network dealing with other matters in the treaty area. A first major step toward an enlarged interest in issues of conservation occurred in 1964 with the adoption of a set of Agreed Measures for the Conservation of Antarctic Fauna and Flora. Emerging in part from scientific research documenting the sensitivity of high-latitude ecosystems to perturbations resulting from human activities, these agreed-upon measures are indicators of a clear turn toward environmental protection as a focus for cooperative measures in the treaty area. These measures constitute a relatively detailed code of conduct, worked out in collaboration with SCAR and designed to govern the activities of scientific personnel operating out of the research stations of the consultative parties with the goal of minimizing environmental impacts arising from the conduct of scientific activities.

With the passage of time, other environmental and resource activities provided the impetus for developments that changed the original 1959 treaty into what we now know as the ATS. The rise of interest in commercial harvesting of Antarctic seals during the 1960s and 1970s led to the adoption in 1972 of the Convention for the Conservation of Antarctic Seals (CCAS), a separate agreement but one that is linked clearly to the Antarctic Treaty and that spells out explicitly a well-defined and critical role for SCAR. CCAS became a stepping-stone to a broader interest in managing human uses of Antarctic marine living resources that grew during the 1970s and led to an agreement in 1980 on the Convention on the Conservation of Antarctic Marine Living Resources (CCAMLR). Driven in part by the expected emergence of large commercial harvests of krill, CCAMLR encompasses all marine systems south of the Antarctic Convergence and is particularly notable for the adoption of a management policy that we now think of as ecosystem-based management. Article II of the convention calls for the "maintenance of the ecological relationships between harvested, dependent, and related populations of . . .

Table 3.1
Antarctic Treaty System Timeline

Year	Event
1822	Antarctica officially discovered
1911–1912	Amundsen and Scott reach the South Pole separately
1914–1917	Shackleton expedition
1900s	7 countries assert jurisdictional claims in Antarctica
1940s	Early proposals for an Antarctic Treaty
1957	Special Committee on Antarctic Research established
1957–1958	International Geophysical Year (IGY)
1959	12 countries sign Antarctic Treaty (AT) on December 1
1958	Special Committee upgraded to the Scientific Committee on Antarctic Research (SCAR)
1961	AT enters into force
1964	Agreed Measures on Conservation of Fauna and Flora
1972	Convention on the Conservation of Antarctic Seals (CCAS) signed
1980	Convention on the Conservation of Antarctic Marine Living Resources (CCAMLR) signed
1980s	New consultative parties admitted to membership
1987	Greenpeace establishes Antarctic base
1988	Convention on the Regulation of Antarctic Mineral Resource Activities (CRAMRA) signed
1991	Review conference becomes an option
1991	Environmental Protocol (EP) to Antarctic Treaty signed
1998	EP enters into force
1998	First meeting of the EP Committee for Environmental Protection
2004	Secretariat of the Antarctic Treaty established
2007–2009	International Polar Year (IPY) 2007–2009
2009	50th anniversary of the signing of the AT
2009	AT has 46 members—28 consultative parties and 18 nonconsultative parties

marine living resources" and requires users to minimize "the risk of changes in the marine ecosystem which are not potentially reversible over two or three decades." The implementation of general provisions of this sort is anything but straightforward. Still, it is fair to characterize this as a progressive approach to management. Experience in this case has played a role in ongoing efforts to develop and operationalize the concept of EBM in other settings.

Even more dramatic developments in the ATS unfolded during the 1980s and 1990s. Responding to the interests of new members and pressured by growing concerns about resource scarcity, the consultative parties launched an effort in the 1980s to devise a supplemental regime governing mineral resource activities in the ATS area. This effort yielded agreement in 1988 on the terms of the Convention on the Regulation of Antarctic Mineral Resource Activities (CRAMRA), an elaborate arrangement designed to allow efforts to explore for and ultimately to develop commercially valuable mineral resources, while at the same time introducing substantial innovations in the management scheme governing these activities. But this remarkable agreement, which had been vigorously opposed from the start by environmental NGOs like the Antarctic and Southern Ocean Coalition (ASOC) and Greenpeace, collapsed without ever entering into force as a result of defections on the part of Australia and France.

The crisis occurring in the wake of the collapse of CRAMRA opened a window of opportunity for a number of governmental and nongovernmental actors to come forward with creative proposals that brought about a watershed change in the ATS (Joyner 1998). The result was an agreement in 1991 on the Protocol on Environmental Protection to the Antarctic Treaty, which entered into force in 1998. This agreement, in the form of an integral part of an expanded Antarctic Treaty rather than an associated measure, calls for the "comprehensive protection of the Antarctic environment . . . [and] designate[s] Antarctica as a natural reserve devoted to peace and science" (Article 2). The emphasis here is on "wilderness and aesthetic values" in contrast to commercial goals as well as on environmental protection in a more conventional sense. Among other provisions, Article 7 reads in its entirety that "[a]ny activity relating to mineral resources, other than scientific research, shall be prohibited." A particularly notable feature of the protocol is the device of addressing specific issues through the addition of annexes to be developed individually and dealing currently with environmental impact assessment, the conservation of fauna and flora, waste disposal and waste management, the prevention of marine pollution, area protection and management, and liability arising from environmental emergencies. Taken together, these developments constitute a striking turn toward the promotion of noncommercial and often intangible values, especially when contrasted with the discourse or vision underlying CRAMRA.

Overall, the ATS has proven remarkably dynamic and highly resilient in its ability to respond to major challenges. This is not just a case of progressive development in the sense of steady and step-by-step fulfillment of an original design, as exemplified in the case of the ozone regime. The ATS has confronted a series of significant challenges and found ways to address them successfully by adding new members along with new elements that were not even implicit in the original design (Repetto 2006). The core of the regime remains unchanged; alleviating political problems through demilitarization and the freezing of jurisdictional claims is just as important today as it was in the 1950s. But the pattern I have called punctuated equilibrium is a prominent feature of the dynamics of this regime. Starting with the effort in the 1960s to emphasize issues of environmental protection, and moving forward during the 1970s and 1980s to introduce EBM as a framework for managing the harvest of living resources and during the 1990s to give a prominent place to "wilderness and aesthetic values," this regime has pioneered innovative thinking regarding the theory and practice of international governance. We cannot predict the future of this regime with precision. But the pattern of challenge and response over the past fifty years adds up to a clear case of punctuated equilibrium.

Analysis: Sources of Punctuated Equilibrium

How can we account for this emergent pattern of punctuated equilibrium in the case of the ATS? Answering this question is not just a matter of explaining the effectiveness of this governance system, although it is possible to make a convincing case that the ATS is one of the most successful members of the growing class of multilateral environmental agreements. Rather, we want to understand the forces that have given rise in this case to the particular pattern of institutional dynamics I call punctuated equilibrium.

Endogenous Factors

At the heart of a cluster of endogenous factors that are conducive to punctuated equilibrium lies a set of features of the ATS pertaining to decision rules, membership criteria, and the legal status of the regime. The decision rules of the regime constitute a force for stasis in the day-to-day operations of the ATS. The Antarctic Treaty itself requires unanimity—not to be confused with consensus—at every step. Ratification on the part of all twelve of the treaty's original signatories was mandatory for the Antarctic Treaty to enter into force, a milestone that occurred in June 1961. Decisions taken under the terms of Article IX (e.g., the 1964 Agreed Measures for the Conservation of Antarctic Fauna and Flora) become effective when approved by all the Antarctic Treaty Consultative Parties (ATCPs). Amendments to the treaty require ratification on the part of all the consultative parties in order to enter into force (Article XII). One implication of these provisions is that there is no need to allow for reservations or objections to specific decisions on the part of individual members, a problem that looms large under the terms of many multilateral environmental agreements. Other elements of the ATS are slightly less demanding or restrictive in these terms. Article XII of CCAMLR, for instance, specifies that decisions by the CCAMLR Commission "on matters of substance shall be taken by consensus," and that "the question of whether a matter is one of substance shall be treated as a matter of substance." This requirement for consensus in decision making regarding living resources serves to highlight the restrictiveness of the Antarctic Treaty's decision rule. Whereas the idea of consensus encompasses situations in which one or more of the parties are willing to go along or abstain rather than blocking the will of the majority, unanimity means that each party must approve a measure explicitly and formally (Breitmeier, Young, and Zürn 2006). The contrast between this arrangement and the more flexible rules under which the ozone regime operates is striking. The designers of the ATS, thinking perhaps of the geopolitical bargain that constitutes the core of the regime, wished to raise the bar high for those advocating changes in the character of the regime.

The ATS has several other endogenous features that cut in the other direction; they make it easier to come to terms on an occasional basis with insistent pressures to make substantial adjustments. One such factor arises from the rules governing membership. At its inception, the 1959 treaty had only twelve members—the ATCPs—all of which had mounted substantial programs of research during the 1957–1958 IGY. Other states are allowed to join the regime by acceding to the treaty, but they cannot attain the status of consultative parties unless they meet the standard set in Article IX(2) of "demonstrating substantial scientific research activity [in Antarctica], such as the establishment of a scientific station or the dispatch of a scientific expedition."[2] Since only consultative parties are allowed to participate in actual decision making during the—now annual—Antarctic Treaty Consultative Meetings (ATCMs), the effect of this arrangement is to restrict the number of parties whose agreement is required to approve significant changes in the character of this regime. The result is a situation in which the requirement of unanimity is stringent but the transaction costs that normally grow—often exponentially—as the number of parties to an agreement rises will be lower than they are in regimes having 180 or more voting members.

Issues regarding membership gave rise to the first major challenge to the regime during the 1970s and 1980s. Increasingly, nonmembers of the regime articulated the view that Antarctica is rightfully a component of the global commons or the common heritage of humankind (much like the moon or outer space) and that the Antarctic regime constituted a club run by a handful of ATCPs whose interests were not coterminous with those of the international community as a whole (Beck 1986; Peterson 1988). Gathering strength especially with the rise of the G77 as a diplomatic force and following the emergence of the campaign for a New International Economic Order (NIEO), this view led to a series of debates during annual sessions of the UN General Assembly aimed at breaking down the clublike character of the ATS and opening the decision-making process of the regime to a more diverse range of perspectives. Eventually, this process produced a decision on the part of the ATCPs to accept a number of additional states as consultative parties as an alternative to bringing the entire regime under the auspices of the United Nations. Today, there are forty-six parties to the Antarctic Treaty, of whom twenty-eight are consultative parties. Persuading twenty-eight parties to agree to substantive changes in the regime is harder than persuading twelve. Even so, it is reasonable to expect that building a coalition of twenty-eight members behind some institutional innovation or reform is a good deal easier than dealing with such issues in a setting in which there are 150 to 200 parties.

The ATS is a stand-alone arrangement that is not part of the UN System and therefore not subject to oversight on the part of the United Nations (Beck 1986). The ATCPs have emphasized the desirability of maintaining cooperative relations with UN bodies throughout the history of the Antarctica regime. Article III of the 1959 treaty itself, for instance, speaks explicitly of the desirability of establishing "cooperative working relations with those Specialized Agencies of the United Nations and other international organizations having a scientific or technical interest in

Antarctica." The links between the ATS and SCAR (and therefore ICSU) are unusually strong for an international environmental governance system. But there is no requirement under the terms of the ATS to consult with any UN bodies or other organizations before taking decisions, much less to obtain UN approval for ATCM recommendations or measures that would constitute important changes in the character of the regime. By contrast, the ozone regime operates under the auspices of UNEP and the climate regime falls within the purview of the UN General Assembly, both bodies that expect to play a role in developments pertaining to these regimes and to participate actively in any efforts to change them significantly. Members of the ATS are free to chart their own course, another feature likely to reduce transaction costs when it comes to mobilizing the political will needed to adopt and implement institutional changes affecting international environmental governance systems.

The Antarctic Treaty contains an explicit provision allowing for the organization of a review of the operation of the treaty. Article XII(2) provides for the organization of a review conference any time after "the expiration of thirty years from the date of entry into force of the present treaty"—after the middle of 1991. A request on the part of any one of the ATCPs is sufficient to activate this provision, and such a conference would allow for a review and potential revision of any aspect of the Antarctic Treaty. No consultative party has called for such a conference, so the Article XII mechanism has never come into play. The general feeling is that holding such a review conference would open Pandora's box and could, as a result, lead to a cacophony of voices proposing institutional reforms that might call into question the regime's core principles of demilitarization and the freezing of jurisdictional claims. Even the more recent additions to the ranks of consultative parties have realized the dangers lurking in such a conference as opposed to a consideration of change in a more focused or piecemeal manner. The consultative parties were hard at work putting the finishing touches on the text of the 1991 Environmental Protocol at the time the option of holding a review conference became operative. The terms of the protocol were hammered out in relatively short order following the collapse of CRAMRA in 1989, a notable achievement given the force of the protocol in articulating an environmental discourse in terms of which to address management issues in the area covered by the ATS. Thus, the consultative parties are perfectly capable of joining forces to agree on substantial changes in important components of the regime, despite the fact that they are leery of resorting to a procedure that could throw open the entire arrangement for reassessment and possible restructuring in constitutive terms.

Given this mixed picture involving features that both facilitate and impede institutional change, it is interesting to note that the ATS also contains provisions that make it relatively easy to implement changes once decisions are made at the political (or legislative) level to adjust the content of the regime. A remarkable feature of this sort arises from the practice of relying on science not only to provide a rationale for the regime as a whole but also to play a pivotal role in operationalizing the terms of any institutional changes that are adopted. Article XI of the Antarctic Treaty emphasizes the role of "[f]reedom of scientific investigation in Antarctica" and calls for "cooperation to that end" as a follow-up to the activities launched in conjunction with the IGY. This theme is not limited to the language of the 1959 treaty. CCAMLR establishes a scientific committee and calls for "cooperation in the field of scientific research in order to extend knowledge of the marine living resources of the Antarctic marine ecosystem." The Environmental Protocol contains numerous references to the role of science, including an obligation on the part of the parties to "promote cooperative programs of scientific, technical and educational value" (Article 6).

What is going on here, and what accounts for the emphasis on science as a core element of the regime, an emphasis that is unique within the class of governance systems based on multilateral environmental agreements? This is where the role of science in the administration of governance systems comes into focus. Scientists and support personnel based at the scientific research stations are the only human residents of Antarctica. The proliferation of these stations in the ATS area resulting from the work of the IGY, coupled with the overarching view that science is by nature a transnational activity that knows no political bounds, provided an opportune mechanism not only for articulating general rules governing human activities in Antarctica but also for designing an elegant monitoring, reporting, and verification (MRV) system that has proved important in ensuring a successful transition of the provisions of the ATS from paper to practice. All members of the regime subscribe to the view that science is an apolitical and cooperative activity that needs to be carried out in a coordinated fashion throughout the treaty area. This has allowed members of the scientific community who move between research stations to play a significant role in promoting the transparency of activities carried out in Antarctica and, in the process, fulfilling the mandate of Article VII of the 1959 treaty dealing with free access for purposes of observation and inspection (NRC 1986). Science does not call the shots when it comes to decisions about proposals to reform or extend the ATS. But it does provide a vehicle that is acceptable to all the consultative parties for implementing a variety of features of the ATS on a day-to-day basis. Once the parties make decisions about changes in the regime, the scientific community provides an effective mechanism to handle the implementation of new arrangements on the ground.

Another endogenous feature of the ATS that is relevant to this account of the pattern of punctuated equilibrium arises from the networked character of the ATS treated as an institutional complex. All the elements of this regime share the core principles of demilitarization, the freezing of jurisdictional claims, and freedom of movement for nationals of the member states throughout the ATS area. Beyond this, the regime provides several tracks for the introduction of changes, including the adoption of agreed-upon measures, the addition of protocols to existing treaties, the development of annexes dealing with particular issues, and the adoption of separate legal agreements dealing with specific issues that are linked tightly to the ATS. The effect of this is to provide flexibility in the selection of policy instruments to address more or less well-defined issues. In the early years, the parties made use of agreed-upon measures developed within the framework of the 1959 treaty as the preferred approach to environmental issues. Many of these measures have since migrated into the Environmental Protocol and become elements of a coherent and legally binding approach to environmental protection. The protocol, in turn, has a series of annexes, dealing with matters such as environmental impact assessment, waste disposal, and liability, which can be adjusted and revised without reopening the substantive provisions of the protocol, much less the Antarctic Treaty itself.

Issues that seem separable, like harvesting living resources in the waters surrounding Antarctica, are subject to management under the terms of distinct but closely linked conventions of their own. The inclusion of CCAMLR constitutes a prominent case in point. A particularly striking episode involving the networked character of the ATS involves the laborious negotiation of CRAMRA, the collapse of this convention following the defection of Australia and France, and the move to pursue almost immediately the negotiation of the Environmental Protocol, which was opened for signature just over two years after the collapse of CRAMRA. The significance of the development of this network stems both from its role in making the ATS an institutional complex and from the role it has played in influencing the discourse in terms of which Antarctic issues are framed and discussed. Whatever else one may say about the performance of the ATS, this development has had the effect of tightening and strengthening the core elements of this regime.

A somewhat different issue involves the interplay between the ATS and other environmental governance systems that have emerged and flourished during the postwar era. Because the ATS seeks to manage a range of issues in a geographically defined space, whereas many other regimes deal with functional issues wherever they occur, it is apparent that interplay will be an important issue confronting the ATS in a number of forms. Sometimes this is a matter of taking advantage of specialized provisions of other arrangements, such as the designation of the Antarctic region as a "special area" under the provisions of Annex I of the International Convention for the Prevention of Pollution from Ships 1973–1978 dealing with the prevention of oil pollution. In other cases, the process features more informal consultations among those representing distinct regimes. A good example is the concern for highly migratory species (e.g., a number of species of birds and whales) that spend time within the ATS area but also leave the area for part of the year and are subject to rules and decision-making procedures associated with other regimes (e.g., the regime created under the terms of the International Convention for the Regulation of Whaling). A more complex situation has arisen in recent years with the establishment and development of the regimes designed to address the depletion of stratospheric ozone and climate change. Biophysical changes in Antarctica loom large in conjunction with ozone depletion and climate change. It is not difficult to integrate research pertaining to the south polar region into the broader stream of research on global environmental change. Research carried out under the auspices of the 2007–2009 International Polar Year, for instance, contributes in a variety of ways to this broader agenda. Still, there are as yet no significant provisions for dealing with interplay between the ATS and evolving regimes addressing ozone depletion and climate change, a topic I return to later in this chapter.

Overall, the ATS has a number of endogenous features that deter change on a day-to-day basis but nonetheless facilitate periodic changes to accommodate major pressures for alterations regarding issues ranging from the decision-making procedures of the regime itself to the growth of interest in new issues like the harvesting of living resources, the development of mining, and the creation of arrangements designed to maintain Antarctica as a natural reserve. What about the role of leadership in this setting? Because the ATS had no secretariat of its own until 2004, some opportunities for leadership of the sort visible in the ozone and climate regimes have not arisen. At least a part of this gap has been filled over the years by key individuals associated with bodies like the British Antarctic Survey (BAS), SCAR, and the Committee of Managers of National Antarctic Programs (COMNAP), a group that plays a central role in the day-to-day administration of the provisions of the ATS.

The contributions of individuals have played an important role in the emergence of specific elements of the ATS. Information now available points to the prominent role Dwight Eisenhower, the U.S. president at the time of the negotiation of the Antarctic Treaty, played in launching the ATS. At a lower level, the efforts of Rüdiger Wolfrum in the development of the Environmental Protocol's Annex on Liability Arising from Environmental Emergencies have made a difference. One of the most remarkable cases of individual leadership in the development of the ATS, the role that Christopher Beeby of New Zealand played in hammering out the terms of CRAMRA, went for naught as the minerals convention imploded and gave way to the development of the Environmental Protocol without ever entering into force.

The situation is less clear with regard to the implementation stage. What we can say is that the ATS has given rise to the emergence of a small but close-knit community of individuals who have come to know one another well and who have become articulate and persistent defenders of this regime within their own governments or organizations like SCAR. Although this group operates on a different plane than the parties to the various components of the ATS, it constitutes an Antarctic community whose members are well-connected and long-term supporters of the Antarctic Treaty System.

Exogenous Factors

What about the role of exogenous factors or forces operating in the world at large in the development of the pattern of punctuated equilibrium that characterizes the dynamics of the ATS? The geopolitical underpinnings of the Antarctic regime continue to loom large, despite major changes in the "high politics" of international society. In the beginning, the role of geopolitics was both transparent and front-and-center. The seven claimant states were not prepared to relinquish their claims to allow Antarctica to be treated as a component of the global commons. In the aftermath of World War II and facing the rise of a bipolar international system, however, the claimants were in no position to prevent access on the part of the superpowers (the United States and the then Soviet Union) to the whole of Antarctica. From their perspective, an international regime acknowledging their claims while providing effective governance in the area seemed to be the best option available. Both the Soviet Union and the United States, by contrast, were determined to assert freedom of movement throughout the area but were attracted to an option that would exempt Antarctica from the tensions of the cold war and from both the costs and the political complications likely to accompany the deployment of military forces in the region. The result was agreement on the core principles of demilitarization and the freezing of jurisdictional claims coupled with a commitment to science as a neutral and mutually acceptable vehicle for the conduct of day-to-day activities in Antarctica.

The end of the cold war and the collapse of the Soviet Union brought about profound changes in the geopolitical equation that led to agreement on the terms of the 1959 treaty. But the appeal of the ATS has remained strong in the broader setting that has emerged during the 1990s and the first decade of the twenty-first century. No one has anything to gain from reconfiguring the core elements of the regime. A revival of jurisdictional claims would only rekindle unwanted tensions, especially among Argentina, Chile, and the United Kingdom, whose claims are overlapping. There is nothing to be gained from the conduct of military activities in the region. Nor are there indigenous peoples striving to articulate claims to land and resources or to political autonomy (as there are in the circumpolar Arctic). Antarctica even lacks appeal to terrorist groups searching for a base of operations from which to launch assaults on the current structure of international society. The result is a coalition of the whole in support of the continuation of the ATS. This is a significant reason why none of the consultative parties has taken the initiative in calling for a review conference of the sort envisioned in Article XII of the Antarctic Treaty.

Several broader political developments have played a role in the development of the pattern of challenge and response that characterizes the dynamics of this regime. The first of these is an outgrowth of the wholesale decolonization of the postwar years, the rise of what we now know as the G77 plus China as an influential bloc or coalition in world affairs, and the tilting of the balance of power in the UN General Assembly toward a new majority often reflecting the views of the G77. Under the leadership of countries like Malaysia, the G77 took early aim at the ATS as a club of advanced industrial countries. Partly an expression of hostility toward the idea that a club of rich countries should be allowed to control a significant component of the global commons, this campaign was motivated also by the hope that Antarctica would prove to be a treasure trove of natural resources whose exploitation could provide material support for supporters of a NIEO. The result was the inclusion throughout a number of years of an Antarctica item on the agenda for the annual meeting of the UN General Assembly along with a war of words about the proper treatment of Antarctica as a part of the global commons (Beck 1986). This challenge subsided when the ATCPs agreed to interpret the provisions of the Antarctic Treaty more liberally to allow a number of additional countries to attain the status of consultative parties. With the granting of consultative-party status to Brazil and India in 1983 and China in 1985, the steam began to go out of this political challenge to the stability of the ATS. The original consultative parties chose to accept a watershed change leading to a new equilibrium involving higher transaction costs arising from the growth in membership over a more chaotic situation in which the ATS as a whole might have foundered and been replaced by a different arrangement under the auspices of the UN (Schelling 1978; Keohane 1984).

Another exogenous development that has presented a challenge to the ATS centers on the rise of the environmental movement and the growth of opposition to any economic development in Antarctica proper. Although they lack any formal or official status in the regime for Antarctica, environmental NGOs have placed a high priority on Antarctica from an early stage and worked hard to exert pressure on the ATCPs. The Antarctic and Southern Ocean Coalition, a group representing a broad spectrum of environmental organizations, has developed a high level of expertise regarding Antarctic matters and has succeeded on many occasions in influencing the tone and the content of debates over matters of policy relating to Antarctica. The fact that Greenpeace has attached high priority to Antarctica and has predictably espoused a more radical position regarding the management of Antarctica has made it possible for ASOC to present itself as the voice of reason regarding environmental issues in the south polar region. Just as the challenge of the G77 pertaining to membership emerged as a focus of debate in the 1970s and 1980s, the debate over the conservation and even preservation of the ecosystems of Antarctica came into focus in the 1980s and 1990s. The turnaround exemplified by the transition from CRAMRA in 1988 to the Environmental Protocol in 1991 represents a watershed change in the life of the ATS. CRAMRA envisioned an elaborate system designed to regulate the exploitation of Antarctica's mineral resources. But the Environmental Protocol not only prohibits "[a]ny activity relating to mineral resources"

(Article 7); it also calls on the parties to "commit themselves to the comprehensive protection of the Antarctic environment and dependent and associated ecosystems and . . . designate Antarctica as a natural reserve" (Article 2). This language falls short of designating all of Antarctica a wilderness area or world park, and the language of the protocol does not preclude future amendments. But the resolution of the challenge of the 1980s and 1990s reflects a remarkable shift toward the preferences of the environmental community.

What are the implications of this shift? It is worth noting, at the outset, that there is nothing in this development that challenges the core principles of the Antarctic Treaty dealing with demilitarization and the freezing of jurisdictional claims in the region. Nor did it annul the provisions of CCAMLR and the seals convention allowing for a well-regulated harvest of marine living resources in the ATS area. The foundation of the regime remains intact, even as a distinct shift in the regime's superstructure has taken place.

Three additional observations will help to elucidate both the sources and the consequences of this pattern of change. The parties to the ATS were able to find ways to safeguard the special status of science, even while adopting the strong commitment of the Environmental Protocol to treating Antarctica as a "natural reserve." The new equilibrium did not run afoul of any entrenched and powerful economic interests in the region. And the reframing of the vision and objectives of the regime reflects the growing influence of a new discourse that has emerged as a force to be reckoned with in the realm of international environmental governance more generally.

Although we sometimes assume that scientific organizations (e.g., SCAR) and environmental organizations (e.g., Greenpeace) should be natural allies with regard to matters like environmental protection in Antarctica, the relationship between the two is characterized by tension in several areas (NRC 1986). Science emanating from the IGY became the vehicle for implementing the provisions of the 1959 treaty. As the Antarctic Treaty Secretariat puts it, "Scientific research is the main activity on the Antarctic continent" (www.ats.aq/e/ats-science.htm). For their part, environmental NGOs (e.g., ASOC) prefer to maintain Antarctica as a pristine protected area or, as some would have it, a world park. They interpret the concept of a "natural reserve" articulated in Article 2 of the Environmental Protocol broadly and worry that scientific research may uncover large untapped deposits of nonrenewable resources in the ATS area. Environmental NGOs have taken strong stands on the design and construction of research stations in Antarctica, the disposal of wastes generated at these stations, and the conduct of field work on the part of scientists pursuing their research throughout the region. It would be a mistake to exaggerate the impact of differences between the two communities; scientists and environmentalists have found common ground regarding activities like the designation of Specially Managed Areas and Specially Protected Areas under the terms of the protocol. Yet tension between these communities is likely to produce an ongoing stream of issues that will surface regularly in venues like the Antarctic Treaty Consultative Meetings.

The story with respect to exogenous economic forces is different and, at least at this stage, favorable to continued cooperation under the provisions of the ATS. The key observation is that (potential) Antarctic resources have not proven attractive to those interested in harvesting living resources or extracting mineral resources on a large scale. Efforts to harvest Antarctic krill have not eventuated in a scramble for resources that would tax the regulatory capacity of CCAMLR. Even more important is the fact that industry did not put up a protracted fight over the transition from CRAMRA, which would have allowed for the development of mineral resources on a regulated basis, to the Environmental Protocol, which prohibits all efforts to develop mineral resources in the ATS area. This could change as a result of the depletion of strategic mineral supplies in other parts of the world. But for now, economic pressures are not taxing the capacity of the ATS to govern human activities in Antarctica, and there are few indications that this situation will change any time soon. The one significant exception to this story involves the steady growth of Antarctic tourism. But not only is tourism heavily regulated through the provisions of the annexes to the Environmental Protocol; tour operators also have a compelling interest in avoiding disruptive activities that would have adverse effects on the appeal of Antarctica to tourists. Recognizing this, the tour operators voluntarily developed a code of conduct governing their activities in Antarctica through the work of the International Association of Antarctic Tour Operators (IAATO). They also have adopted a cooperative stance in their interactions with the various elements of the Antarctic regime. Overall, exogenous economic forces have not put heavy pressure on the ATS to accommodate forces advocating a growth of industrial activities in and around Antarctica.

The rise of a new discourse emphasizing sustainability and highlighting the idea of ecosystem-based management as an alternative to the traditional focus on maximum sustained yields (MSY) has played a major role in the evolution of the ATS and especially the dramatic shift from the prospect of development implicit in CRAMRA to the spirit of preservation underlying the Environmental Protocol (Larkin 1977). This shift constitutes a worldwide movement; it is in no way specific to the regime for Antarctica (Dryzek 1997). Yet Antarctica has provided an attractive setting in which to experiment with various approaches to EBM without triggering passionate reactions on the part of affected resource users. I have commented already on the language in CCAMLR calling for a shift to EBM as a basis for fisheries management. The protocol goes even further, emphasizing the importance of "wilderness and aesthetic values" and calling for the treatment of the whole ATS area as a "natural reserve devoted to peace and science." The process of moving these concerns from paper to practice is not easy or straightforward. CCAMLR continues to wrestle with the operational meaning of the injunction to minimize "the risk of changes in the marine ecosystem which are not potentially reversible over two or three decades" (Article II). Assuming that human activities including scientific research continue to take place in Antarctica on a sizable scale, the effort to strike a balance between "wilderness and aesthetic values" and the benefits to be derived from somewhat messy human activities (e.g., drilling ice cores to shed light on the history of the Earth's climate system) will constitute an ongoing problem. Nevertheless, the rise of this new discourse, brought about largely through the efforts of nonstate actors, is already shaping the Antarctic agenda; it is likely to become even more influential as we move toward a growing emphasis on sustainability worldwide.

This brings me to a final exogenous factor destined to loom large in future thinking about south polar region governance. Global environmental changes are not products of human activities taking place in the high latitudes. But the impacts of these changes are already being felt in Antarctica. The dramatic seasonal thinning of stratospheric ozone was documented definitively for the first time as a result of observations made in Antarctica. The most severe and protracted annual depletions of stratospheric ozone (the ozone hole) and, as a result, increased UVB radiation reaching the Earth's surface occur in the south polar region. The impacts of climate change are more dramatic in the polar regions than elsewhere. Although impacts to date are more extensive and intensive in the high latitudes of the northern hemisphere, Antarctic impacts are not far behind. The most visible effects of climate change in the ATS area will center on the disintegration of ice shelves. The Larsen B shelf, covering an area larger than the state of Rhode Island, disintegrated during the austral summer of 2002. Observers have now shifted their attention to the fate of the West Antarctic Ice Sheet. Were this large mass of ice to disintegrate, it would raise sea levels worldwide by some five meters, causing immense damage to human welfare around the world in the process. The ATS is not in a position to take actions that will prevent the occurrence of climate change and stabilize conditions in Antarctica in the face of major changes in the Earth's climate system. But it is important to recognize the role of climate change as an exogenous force and to consider the role of Antarctica in this context. One interesting opportunity here has to do with the conduct of research in Antarctica that is relevant to the search for knowledge regarding global systems like the climate system. A project is currently under way to drill an ice core that will allow scientists to make observations regarding the climate system that go back 1.5 million years. The role of the International Polar Year (IPY) of 2007–2009 administered by ICSU and structured in such a way as to emphasize links between polar processes and global phenomena has produced constructive efforts to take into account links between polar and global processes.[3]

Endogenous-Exogenous Alignment

Given this array of endogenous and exogenous factors affecting the ATS, what can we say about interactions among them? There is a fundamental and essential connection between the forces of geopolitics and the central features of this regime. The cold war conditions that made demilitarization, the freezing of jurisdictional claims, and the turn to science as a vehicle for realizing cooperation are no longer with us. Yet no party to the regime is interested or willing to take the risk of undermining the basic bargain articulated in the 1959 treaty and still central to the character of the regime today. The Environmental Protocol with its provisions for the Committee for Environmental Protection that performs a range of functions relating to the protection of Antarctic ecosystems adds another dimension to this foundational arrangement. Efforts to protect ecosystems are joined to the conduct of scientific research as a vehicle for ensuring that the basic bargain of this governance system remains intact. This is not ultimately a matter of environmental governance. Yet the geopolitical foundation of the ATS emerged at an early stage as a condition conducive to the development of an array of innovative environmental arrangements. There is every reason to expect this situation to remain central to the ATS over the course of the foreseeable future.

The predominant pattern in the interaction between endogenous and exogenous factors in the ATS involves an episodic drift away from equilibrium, coupled with periodic and substantial changes needed to restore the internal-external balance. The cases of membership and the harvesting of living resources fit this pattern clearly. When pressures built up during the 1970s and 1980s to do something about the character of the regime as a rich man's club, the ATCPs chose to let in a number of new consultative parties rather than see the ATS get absorbed into the UN System. When serious interest in harvesting living resources in and around Antarctica arose during the 1970s, the ATS responded with the addition of the seals convention and, more important, CCAMLR as separate but closely connected governance systems.

A more complex case involves the changes occurring in the 1980s and 1990s centered on efforts to negotiate the terms of CRAMRA and, shortly thereafter, the terms of the Environmental Protocol. My interpretation of the record regarding these changes is that developing countries pushing the idea of the NIEO—more so than major corporations engaged in extractive industries—constituted the driving force behind the negotiations relating to mineral activities that culminated in the provisions of CRAMRA in 1988. Domestic politics leading to the defection (for different reasons) of Australia and France sank CRAMRA, since the convention could not enter into force without the participation of all the ATCPs. The convention appears to have fallen victim to electoral politics in Australia, and France apparently seized this opportunity to burnish its somewhat lackluster environmental record with a policy shift that would not be costly in terms of domestic politics. Why did the developing countries among the ATCPs (e.g., Brazil, China, and India) accept the demise of CRAMRA without making a fuss? In part, the actions of Australia and France presented them with a fait accompli; there was no way to resurrect CRAMRA so long as these key states were unwilling to agree to such a move. But partly, it appears that the lack of interest in Antarctic mineral resources on the part of major corporate players giving rise to a growing sense that these resources would not become a commercial bonanza at any time in the foreseeable future played a critical role in the waning of support for CRAMRA.

This set the stage for the emergence of the Environmental Protocol and with it a restoration of the balance between internal and external forces. But why did the protocol emerge so quickly following the collapse of CRAMRA, and why did the parties agree to the broad provisions included in the protocol? The answer to these questions lies in the combination of the growing influence of a new discourse in terms of which to frame Antarctic issues, the decline of expectations on the part of the developing countries regarding the prospect of reaping material gains from the exploitation of Antarctic mineral resources, and the desire of Australia and France—the CRAMRA defectors—to show their concern for good governance in the ATS area. Only the United States seemed reluctant to join the bandwagon in favor of the protocol, and even American concerns seem to have had more to do with a general leeriness regarding the rapid pace of change than with serious opposition to the contents of the protocol. In the end, the adoption of the protocol fit into the preexisting structure of the ATS without requiring any changes of a material sort. There is nothing in the protocol that calls into question the bedrock proposition articulated in the preamble to the 1959 treaty asserting that "it is in the interest of all mankind that Antarctica shall continue forever to be used exclusively for peaceful purposes and shall not become the scene or object of international discord." If anything, the protocol reinforces this proposition by adding activities directed toward environmental protection to the conduct of scientific research as vehicles for realizing the vision of the preamble. The adoption of the Environmental Protocol also dovetails with the propensity of the ATS to make major changes at relatively long intervals in contrast to smaller changes on a regular basis. This episode exemplifies the pattern of punctuated equilibrium that is the dominant mode of institutional change in this regime.

Along the way, the ATS has acquired organizational mechanisms needed to address the tasks the ATCPs have assigned to it. The 1959 treaty made no provisions for such mechanisms. One of the consultative parties volunteered to organize each ATCM; the only major outcomes took the form of agreed-upon measures adopted under the terms of Article IX and entrusted to the individual member states to implement with regard to the activities of their nationals in Antarctica. But CCAMLR created the Commission for the Conservation of Antarctic Marine Living Resources to handle regulatory matters relating to the harvest of living resources. The Environmental Protocol established the Committee for Environmental Protection to manage an array of regulatory provisions set forth for the most part in the various annexes to the protocol. In 2004, the ATCPs finally acknowledged the realities of the situation and created a small Antarctic Treaty Secretariat located in Buenos Aires, Argentina. Taken together, these organizational arrangements have added to the capacity of the ATS to handle a variety of tasks that have emerged over the course of fifty years since the signing of the Antarctic Treaty. One consequence of these developments is to upgrade the ability of the ATS to deal effectively with a raft of day-to-day matters and, increasingly, to play a role in framing the issues and setting the agenda for consideration at the ATCMs. The ATS remains a relatively lightly administered regime. But the development of administrative capacity has gone some way toward making the regime an actor in its own right rather than limiting its role to the provision of a forum in which member states can address Antarctic issues in an orderly fashion. On balance, it is reasonable to expect that this development will enhance the capacity of the ATS to respond effectively to new challenges likely to arise in the foreseeable future.

Forecast: The Road Ahead

The ATS is one of the most successful international regimes of the postwar era. Antarctica has not experienced any movement toward militarization. The provisions of Article IV of the 1959 treaty have curbed the growth of jurisdictional claims, while offering reassurance that nothing in this arrangement constitutes a renunciation of "previously asserted rights of or claims to territorial sovereignty in Antarctica." Several ATS members (e.g., Argentina, Australia, Chile, and the UK) have made claims recently under the provisions of Article 76 of the 1982 UN Convention on the Law of the Sea to areas of the continental shelf extending beyond two hundred nautical miles from the coast of Antarctica.[4] Nonetheless, it seems accurate to conclude that the jurisdictional claims that existed prior to the adoption of the treaty have atrophied; it is hard to imagine any serious effort to revive terrestrial claims in the foreseeable future. The original treaty has morphed into the Antarctic Treaty System, an institutional complex including additional conventions, agreed-upon measures, and the Environmental Protocol to the original 1959 treaty. The system now includes some of the most innovative arrangements in the entire field of international environmental governance, including the ecosystem-based management approach to marine living resources set forth in CCAMLR and the systems for Specially Managed Areas and Specially Protected Areas along with the liability arrangements set forth in the annexes to the Environmental Protocol. It is relatively easy to establish the causal significance of many of these arrangements. It is hard to imagine jurisdictional claims atrophying in the absence of the regime, practices relating to protected areas and waste disposal follow the prescriptions of the regime, and compliance with the highly restrictive regulations governing human activities in Antarctica is high. Altogether, the ATS stands out as a success story in the realm of international governance.

What lies ahead for the ATS? It is unlikely that the geopolitical core of the regime will disintegrate in the foreseeable future. The key prescription spelled out in Article I of the 1959 treaty that "Antarctica shall be used for peaceful purposes only" has evolved into an article of faith. No party has taken steps that challenge the Article I prohibition on "any measures of a military nature, such as the establishment of military bases and fortifications, the carrying out of military maneuvers, as well as the testing of any type of weapons." Nor does any party have interests that would lead to irresistible incentives to ignore or even to chip away at this blanket prohibition. Because the terrestrial claims of the claimant states have receded into the background, it seems increasingly reasonable to treat Antarctica as part of the global commons. Moreover, to the extent that Antarctica produces values that are nonrival or nonsubtractable (e.g., scientific knowledge or amenities associated with maintaining Antarctica as a "natural reserve"), it is possible that the area can remain open to a range of uses, including scientific research and regulated tourism, without running into the problem widely known as the tragedy of the commons (Baden and Noonan 1998).

At a somewhat more mundane level, it is possible to identify a number of strains that have arisen in the range of human activities taking place in the ATS area and that will continue to arise in the future. Many environmentalists are critical of the practices of members of the scientific community with regard to such matters as the design of scientific stations and the treatment of wastes. Scientists are often irritated by the expectations of tour operators regarding matters like the provision of weather forecasts, the opening of research stations to visitors, and preparedness to mount rescue operations in the event of emergencies. Tour operators and tourists sometimes chafe at the increasingly strict regulations regarding landings of ship-based tourists on Antarctic shores. Although most Antarctic activities now take place under highly restrictive regulations, there remains the chance that environmentally destructive accidents will occur. The 1989 sinking of the *Bahia Paraiso,* an Argentine polar transport carrying some 600,000 liters of petroleum products, in the waters adjacent to the American Palmer Station provides a dramatic example. Although less disturbing, the sinking of the cruise ship *Explorer* in 2007 makes it clear that such accidents can occur at any time. We should not dismiss such strains lightly. Yet they are all normal occurrences in situations involving a number of distinct actors whose interests are not identical, even though they share a fundamental interest in protecting the integrity of a region like Antarctica. Such strains will not disappear; they are likely to flare up from time to time over specific issues. But they will not pose any profound threats to the effectiveness of the Antarctic Treaty System.

The most significant cloud on the horizon today is global environmental change in its various manifestations. The annual thinning of stratospheric ozone is most acute over Antarctica. The phasing out of CFCs and other ozone-depleting substances (ODSs) under the terms of the Montreal Protocol has resulted in a major decline in production and consumption of these chemicals and seems likely to reduce or even eliminate the annual ozone hole in due course. Because of the long residency of ODSs in the atmosphere, however, there is as yet no clear trend regarding reductions in the size of the ozone hole (see figure 3.1). Even more far-reaching are the expected impacts of climate change on Antarctica. Major ice shelves (e.g., the Larsen B Shelf) have already disintegrated, probably attributable in part to the effects of climate change; attention is now focused on the fate of the West Antarctic Ice Sheet. Physical changes of this nature can pose threats to the well-being of Antarctic fauna, including various species of penguins as well as migratory species (e.g., whales) that are seasonal residents of the ATS area. Assuming the projections of the scientific community—in such forms as the assessment reports of the Intergovernmental Panel on Climate Change—are correct, the impacts of global changes on Antarctica are destined to become more severe in the next several decades (IPCC 2007).

A striking characteristic of these changes from the perspective of the ATS is that they are highly asymmetric. Although they can and almost inevitably will have major impacts on Antarctica, they are driven by anthropogenic activities taking place outside Antarctica and well beyond the capacity of the ATS to regulate. There is no prospect of expanding the scope of the ATS to cover the human activities causing the ozone hole, the disintegration of ice sheets, or the deteriorating conditions affecting the welfare of a variety of species. The one ray of hope in this picture arises from the fact that most of the states responsible for problems like climate change are also ATCPs and that evidence arising from research in and on Antarctica may play a role in pushing global environmental change to the top of the international policy agenda. Research conducted in Antarctica was highly influential in documenting the depletion of stratospheric ozone and in producing decisive evidence regarding the causal mechanism underlying this phenomenon. Dramatic changes in ice sheets, observed in Antarctica and in Greenland, are helping to

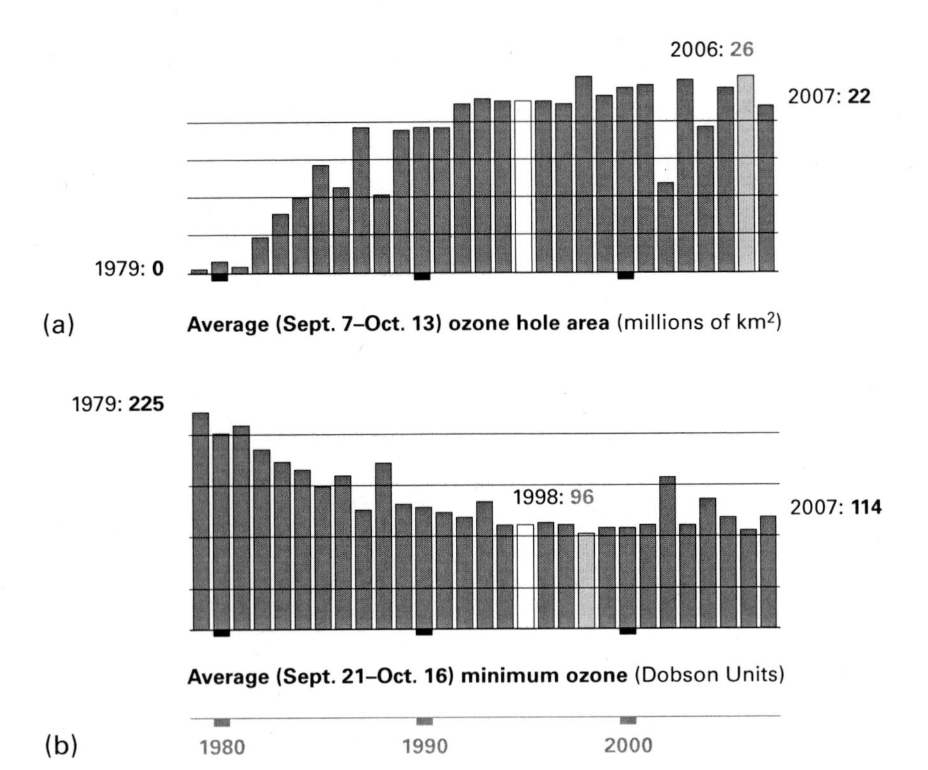

Figure 3.1
Ozone hole annual record since 1979. *Note*: No data were acquired during the 1995 season. *Source*: NASA Ozone Hole Watch, Goddard Space Flight Center, http://ozonewatch.gsfc.nasa.gov/statistics/meteorology_annual.png.

highlight the consequences of climate change and to make them understandable to members of the attentive public. The recent International Polar Year, beginning in the spring of 2007 and lasting through the spring of 2009, provided an important opportunity in this realm. The 1957–1958 IGY played a role in preparing the ground for the adoption of the 1959 Antarctic Treaty. The results of the 2007–2009 IPY may prove helpful in preparing the ground for innovative responses to global issues, like the provisions of a new agreement dealing with climate change to take effect following the expiration of the Kyoto Protocol's commitments at the end of 2012.

Conclusion

A feature of the Antarctic Treaty System that sets it apart from most other international environmental regimes is that its remit extends to matters of high politics and security, including a ban on military activities in Antarctica and the freezing of jurisdictional claims pertaining to the region. This nonenvironmental base has provided a foundation on which an extensive and innovative environmental governance system has arisen. The ATS is a success story in the field of international environmental governance. It has been able to adapt to changing circumstances and to add components that have given rise to an institutional complex extending well beyond the content of the original Antarctic Treaty. But unlike the pattern of progressive development that marks the ozone regime, the ATS exhibits the pattern of institutional dynamics I call punctuated equilibrium. The regime has a number of features that act as deterrents to change on a continuous or routine basis. But it also has a demonstrated capacity to make substantial changes to address major challenges on an episodic basis. Striking examples include the admission of new ATCPs as a means of defusing criticism of the regime's clublike character, the development of ecosystem-based management as an approach to regulating harvests of living resources, and the transition from the vision of regulated mining embodied in CRAMRA to the commitment to protecting Antarctica as a "natural reserve" spelled out in the Environmental Protocol. Will the ATS find a way to address the looming challenge associated with the onset of global environmental changes? It would not make sense to expect the regime for Antarctica to become a vehicle for coming to terms with overarching problems such as climate change. But it would not be surprising to find this arrangement emerging as one element in a broader strategy aimed at coming to terms with some of the grand challenges of our era.

What accounts for the occurrence of punctuated equilibrium as the dominant pattern of institutional change in the case of Antarctica? I argue that this pattern is the result of a combination of endogenous factors that make it hard to address pressures for change on a gradual or piecemeal basis and exogenous factors that have altered the cast of players and the political forces affecting the ATS area in significant ways. The result is a pattern in which pressures build up with regard to issues like membership or environmental protection, the regime fails to respond effectively until these pressures become relatively acute, but eventually the regime introduces major adjustments (e.g., reinterpreting the criteria for consultative-party status) that make it possible to meet the challenge without transforming the character of the regime in the process. In the terminology introduced in chapter 1, this is a situation in which robustness periodically gives way to resilience, and the regime has proven sufficiently resilient to perform its basic functions effectively without crossing a threshold that would lead to change of a constitutive nature. The secret to success in this case has not been a matter of toughing out external pressures without significant adjustments in institutional capacity. Rather, the ATS has shown a remarkable capacity to respond once external pressures become severe, while keeping its defining features intact.

4

Arrested Development: The Climate Regime

Overview: The Big Picture

Life on Earth is dependent on the presence of greenhouse gases (GHGs) in the planet's atmosphere. Dramatic reductions in the concentration of GHGs would lower temperatures to a point where the Earth would become uninhabitable. Sharp increases would raise temperatures to a level that would make contemporary societies nonviable and eventually threaten life on the planet altogether. The important thing with regard to climate, therefore, is to maintain a proper balance in the concentration of these gases. Over the past half million years at least, natural forces have proven remarkably effective in maintaining stability in levels of GHGs in the atmosphere. Atmospheric concentrations of carbon dioxide, the most important greenhouse gas, have fluctuated over a range of about 180 to 280 parts per million (ppm). But the onset of the industrial revolution and the resultant rise in human consumption of fossil fuels have broken this pattern. The current concentration of carbon dioxide is about 387ppm; it is rising at a rate of 2–2.5ppm per year (Steffen et al. 2004). The consensus within the scientific community is that this dramatic departure from the long-term pattern is attributable largely to anthropogenic forces (IPCC 2007). Many uncertainties remain concerning the dynamics of the Earth's climate system and the likely impacts of rising atmospheric concentrations of CO_2 and other greenhouse gases. But almost everyone now agrees that the Earth has entered a no-analog state with regard to climate and that it is critical to prevent concentrations of carbon dioxide (not to mention other GHGs) from rising above two times preindustrial levels, or about 540ppm (Steffen et al. 2004). There is an emerging consensus among scientists and policy makers who focus on climate change that even this level is too high and that we should be aiming to reach equilibrium at 450ppm. A rapidly growing segment of the scientific community now asserts that stabilizing at 350ppm is a better target.

One consequence of this development is that climate change has become the poster issue for those concerned with global environmental change and with the emergence of a new body of knowledge known as Earth system science. Climate change is systemic in character; changes attributable to occurrences anywhere on Earth will have planetary effects in this realm. The Earth's climate system has become a paradigmatic example of a coupled socioecological system requiring greatly increased collaboration between natural scientists and social scientists endeavoring to understand the behavior of this system and to reach conclusions about policy measures needed to address the problem of climate change (Flannery 2005; Linden 2006). Global environmental change is not limited to the problem of climate change. But many thoughtful observers believe that coping with climate change—preventing profound disruptions and adapting to major changes—is emerging as one of a handful of issues that will define the era unfolding in the twenty-first century, which increasingly is being described as the Anthropocene (Stern 2007).

There are significant similarities between the cases of climate change and stratospheric ozone depletion. As with ozone depletion, we can approach climate change either as an externality or as a matter of overuse of a free factor of production, and we need to reckon with the prospect of free-ridership in devising strategies to control climate change or adapt to its consequences. The temptation to use the atmosphere as a repository for wastes or residuals available free of charge is particularly strong in this connection. Efforts to date to address the problem of climate change bear a marked resemblance to the response to stratospheric ozone depletion. The members of international society started with a framework convention—the 1992 UN Framework Convention on Climate Change (UNFCCC)—and progressed to a more substantive protocol—the 1997 Kyoto Protocol—with the hope that these steps would set in motion a process of progressive development much like the process that has occurred in the case of ozone. But, so far at least, the results have diverged dramatically in the two cases. Whereas production and consumption of ozone-depleting substances (ODSs) have fallen dramatically in the period since the signing of the Montreal Protocol, worldwide emissions of greenhouse gases have continued to rise steadily. There is as yet no end in sight.

How can we account for this difference? This chapter explores this question in depth. But it is helpful as a point of departure to observe that the Earth's climate system has no parallel as a complex and dynamic system. Leaving aside anthropogenic drivers, biophysical forces ensure that the Earth's climate system is highly dynamic (Alley 2000; Mayewski and White 2002). The role of greenhouse gases is merely one of a host of factors that serve as determinants of the behavior of this system. A striking characteristic of the system is the occurrence of a number of distinct cycles (e.g., Milankovitch cycles, Dansgaard-Oeschger cycles, cycles in sunspot activity, and the North Atlantic Oscillation) that operate independently of one another but also interact to produce sharp climatic swings between ice ages and more moderate periods lasting for thousands of years. Not only does this give rise to nonlinear processes and patterns of change that differ dramatically from one another but it also produces thresholds and tipping points that can lead to abrupt changes in the climate system as well as emergent properties that are hard to account for in terms of analyses of the processes leading up to them. From this perspective, sharp increases in concentrations of greenhouse gases associated with human actions are properly thought of as adding an additional element to a complex stew of forces that can and do trigger changes that are often nonlinear, sometimes abrupt, and frequently irreversible. Given the unusually stable and benign state of the climate system during the Holocene, a period encompassing the past ten thousand years, future changes can be expected to be both unfamiliar and nasty from the perspective of the Earth's human residents. Yet far-reaching climate change is hardly unprecedented in the history of the Earth (Alley 2000; Mayewski and White 2002).

A prominent feature of the resultant situation is a high level of uncertainty about what lies in store for us. Efforts to analyze the Earth's climate system in holistic terms cannot proceed via the familiar practice of developing deterministic models and comparing their results with empirical observations. The best we can hope for is the development of simulations, in such forms as general circulation models (GCMs), together with inductive assessments of past changes in the Earth's climate system (e.g., the onset and decline of ice ages). These methods have contributed greatly to our understanding of the Earth's climate system; we know a good deal about the sensitivity of this system to changes in the value of key parameters as well as the conditions that trigger abrupt changes. But as leading climate modelers are quick to observe, this does not allow us to forecast with any confidence the future behavior of the Earth's climate system (Bolin 1997). All those seeking to address the problem of climate change must be prepared to deal with uncertainty and to make needed adjustments in both goals and policy instruments as new information becomes available.

This explains the difficulty in articulating a clear-cut policy goal to guide the development of the climate regime. The 1992 UN Framework Convention on Climate Change describes the "ultimate goal" of the climate regime as the "stabilization of greenhouse gas concentrations in the atmosphere at a level that would prevent dangerous anthropogenic interference with the climate system" (UNFCCC Art. 2). But neither the convention nor the more substantive 1997 Kyoto Protocol (KP) translates this goal into an operational target. Whereas the ozone regime is fundamentally a prohibition regime starting with sharp reductions in the production and consumption of ODSs and, in most cases, evolving into a complete phaseout of the use of these chemicals, the climate regime calls for maintaining concentrations of GHGs in the atmosphere at a level that will prevent human actions from disrupting the climate system. Given the nature of the problem, vagueness regarding goals is understandable. This vagueness has become a source of problems plaguing efforts to move the climate regime from paper to practice and, in the process, to come up with the quantitative targets and timetables needed to implement the provisions of the regime.

The need to treat the climate system in complex and dynamic terms also helps to explain the adoption of the familiar framework convention/protocol approach to regime building in this case. There is much to be said in a case of this sort for an approach that starts with a general framework and assumes that more substantive provisions can and will be added as the need arises. Yet this understandable strategy is also an important factor contributing to the occurrence of arrested development in the case of the climate regime. In this chapter, I argue that there is a substantial mismatch between the nature of the problem of climate change and the character of the regime created to deal with it. The problem features a complex and dynamic system whose behavior we cannot predict with any confidence but that acts in a manner that can and sometimes does produce changes that are nonlinear, abrupt, irreversible, and nasty, at least from a human point of view. The regime, on the other hand, is based on the assumption—implicitly if not explicitly—that an incremental, step-by-step strategy will lead eventually to the development of a governance system capable of dealing effectively with the problem of climate change. There is nothing wrong with strategies of this sort in generic terms. The difficulty with regard to climate change lies in the fact that there is a mismatch between the nature of the problem to be solved and the character of the governance system created to solve it. Whether key features of international society make mismatches of this sort unavoidable in dealing with problems like climate change is an important question to which I return later in the chapter. But whatever the answer to this question, it does not alter the explanation for the occurrence of arrested development in this case.

Facts: A Brief History of the Climate Regime

Scientists have understood the role that greenhouse gases play in the Earth's climate system, at least in general terms, since the late nineteenth century. The Swedish chemist Svante Arrhenius even focused explicitly on temperature changes arising from increases or decreases in concentrations of GHGs in the atmosphere. He made projections regarding likely increases in temperature associated with rising concentrations of GHGs that parallel the best estimates of the Intergovernmental Panel on Climate Change (IPCC) even today. But he failed to understand the speed with which the buildup of GHGs could occur. He and others contemplating this issue consequently thought of climate change as a novelty, something that would occur in the far distant future, if at all. Climate change did not emerge at that stage as an issue deserving the attention of policy makers at any level (Flannery 2005).

All this changed dramatically with the identification during the 1950s and 1960s of what we now know as the Keeling Curve. Based on continuous monitoring starting in 1957, this curve demonstrates decisively that levels of carbon dioxide in the atmosphere not only fluctuate on the basis of an annual cycle but also exhibit a clear upward secular trend over time. Recent measurements show that the rate of increase has risen in recent decades. Concentrations are now more than a third higher than preindustrial levels. We may already be committed to a doubling of GHGs in the atmosphere relative to preindustrial levels. It was inevitable, under the circumstances, that climate change and variability would become a prime focus of scientific attention and subsequently an issue of increasing concern to policy makers (Dessler and Parson 2006).

As table 4.1 shows, some time passed before climate change became a prominent topic, even within the scientific community. Scientific awareness of climate change and variability actually predates concern about threats to stratospheric ozone. But the complexity of the Earth's climate system has made it hard to arrive at clear-cut conclusions about the consequences of rising levels of greenhouse gases in the atmosphere. Based

Table 4.1
Climate Regime Timeline

1890s	Arrhenius makes calculations on the role of greenhouse gases
1957	Monitoring starts that leads to the Keeling Curve
1979	First World Climate Conference
1985	Villach Conference
1988	UNEP/WMO establish IPCC
1988	Toronto Conference on CO_2 in the atmosphere
1990	Intergovernmental Negotiating Committee (INC) formed
1991	IPCC First Assessment Report
1992	UNFCCC signed at UNCED
1994	UNFCCC enters into force
1995	COP 1—Berlin Mandate
1995	IPCC Second Assessment Report
1997	COP 3—Kyoto Protocol signed
2001	IPCC Third Assessment Report
2001	U.S. renounces Kyoto Protocol
2001	Marrakech Accords to implement the Kyoto Protocol adopted
2005	Kyoto Protocol enters into force
2005	COP 11 and MOP 1 meet in Montreal
2007	IPCC Fourth Assessment Report
2007	Bali Roadmap adopted at COP 13
2008–12	First commitment period under Kyoto Protocol
2009	COP 15 meets in Copenhagen
2013–	Post-Kyoto climate regime needed

on a more general assessment of climate cycles, many scientists contemplating the Earth's climate system during the 1960s and 1970s thought that a more serious concern was the prospect of reaching a tipping point that would trigger the onset of a colder period or even a new ice age. From this perspective, any warming associated with rising concentrations of GHGs in the atmosphere would either act as a brake on movement toward a new ice age or be irrelevant given the larger forces at work.

The late 1970s and especially the 1980s brought marked changes resulting from the work of a series of high-level conferences and finally a decision on the part of the World Meteorological Organization (WMO) and the UN Environment Programme to create the Intergovernmental Panel on Climate Change in 1988. The idea of anthropogenic forcing via emissions of GHGs took root and flourished as a consequence both of continuing data collection and of a sequence of conferences including the First World Climate Conference in 1979, the Villach Conference in 1985, and the Toronto Conference on CO_2 in the Atmosphere in 1988. With the creation of the IPCC, a new practice now known as scientific assessment came about (Mitchell et al. 2006). Assessment in this sense features an effort on the part of the scientific community to synthesize all available scientific findings regarding a more or less well-defined topic in order to take stock of what we currently know about the topic and to identify areas where additional research is needed (Farrell and Jäger 2006; Mitchell et al. 2006). Coupled with the occurrence of extreme temperatures during the summer of 1988, this process proved sufficient to move the issue of climate change to a relatively prominent position on the environmental policy agenda by the end of the decade.

The result was a sequence of policy initiatives pertaining to climate change that seemed promising, at least during the early stages of the process. Mandated by a resolution of the UN General Assembly (UNGA Res. 212[XLV]), the Intergovernmental Negotiating Committee (INC) began work during 1991 on the text of an international convention on climate change. Driven by the desire to have a product available in time to be opened for signature at the June 1992 UN Conference on Environment and Development (UNCED) and influenced by the recent experience with stratospheric ozone, the INC opted for a framework convention/protocol approach to addressing what was now widely regarded as the problem of climate change. This effort proved successful. Opened for signature at Rio, the UN Framework Convention on Climate Change entered into force during 1994. The first meeting of the Conference of the Parties (COP 1) took place in Berlin during 1995. Some observers interpreted this sequence of events as evidence of a prompt start in addressing the problem of climate change.

At COP 1, the parties recognized explicitly that the provisions of the UNFCCC would not suffice to come to terms with the problem of climate change; they agreed to create the Ad hoc Group on the Berlin Mandate (AGBM) as a means of laying the groundwork for a substantive protocol somewhat similar in character to the Montreal Protocol in the case of ozone. This process eventuated at COP 3 in 1997 in the signing of the Kyoto Protocol, an agreement calling on the advanced industrial countries—designated as the Annex 1 countries under the terms of the

UNFCCC—to cut their emissions of greenhouse gases by a collective average of 5 percent relative to 1990 levels by the end of a first commitment period running from 2008 through 2012.

At this stage, however, problems began to surface. The Kyoto Protocol did not enter into force until 2005. Although the United States had signed the protocol in 1997, the newly installed Bush administration announced early in 2001 that the United States would withdraw from the protocol and refuse to accept any obligation to comply with its provisions and especially the provision regarding targets and timetables for the reduction of GHG emissions. To meet the conditions governing entry into force of the protocol—fifty-five parties and at least 55 percent of 1990 CO_2 emissions on the part of Annex 1 countries—the remaining signatories initiated a process designed to save the protocol and bring it into force by increasing the availability of flexibility mechanisms and offering concessions to key parties, especially Russia and the other countries with economies in transition. The result of this process was the adoption in 2001 of the Marrakech Accords, a kind of regulatory guide for the actions of individual member countries in their efforts to implement the provisions of the protocol and to meet their commitments to reducing emissions of GHGs. The Kyoto Protocol would not have entered into force at all in the absence of the Marrakech Accords. Still, the price of success in achieving this result was high. The parties accepted provisions regarding flexibility mechanisms that have since proved troublesome; they acceded to the arguments of Russia and other countries with economies in transition regarding the value of involuntary reductions of emissions during the 1990s, and they moved forward without any plan for persuading the United States to enter the regime. Under the Marrakech Accords, the de facto collective commitment on the part of remaining Annex 1 signatories to reductions of GHG emissions was watered down substantially.

The result, predictably enough, has been a steady rise in worldwide emissions of greenhouse gases.[1] By 2004, Japanese emissions had risen 5.3 percent relative to 1990 levels, and the comparable figures for the United States and Canada were 14.4 percent and 21.7 percent, respectively. Only the EU-15 had actually reduced emissions. Even so, the reduction by 2006 relative to 1990 levels was only 2.2 percent, substantially less than the commitment to reduce emissions by 8 percent by the end of the first commitment period under the Kyoto Protocol. Concurrently, emissions on the part of non–Annex 1 countries were rising, rapidly in some cases. Chinese emissions, for instance, rose by 50 percent in the decade between 1994 and 2004. China's current growth rate in GHG emissions is well below the rate of growth of China's GDP, testimony to serious efforts on the part of Chinese policy makers to increase energy efficiency. Still, the best estimates indicate a growth rate in Chinese emissions of around 4 percent per annum (Congressional Research Service 2008). China has surpassed the United States as the world leader in GHG emissions. Together, China and the United States now account for approximately 40 percent of the annual emissions of carbon dioxide and some 35 percent of all GHGs (Congressional Reference Service 2008). Adding together the emissions of the non–Annex 1 countries and the nonsignatories leads to the conclusion that half or more of current worldwide emissions of GHGs are not subject to any obligations regarding reductions under the provisions of the climate regime.

A particularly worrisome concern today arises from the fact that the most recent IPCC assessment—the 2007 Fourth Assessment Report—indicates that cuts of 60 to 80 percent in current emissions will be needed by 2050 just to prevent concentrations of GHGs in the Earth's atmosphere from rising above 450ppm. Those who defend the Kyoto Protocol naturally treat this effort to address the problem of climate change as a work in progress. They see the protocol merely as a first step and believe that it is possible to ratchet up reduction obligations over time in a manner resembling the progressive development of the ozone regime. But this line of thinking does not inspire confidence. Taken together, the rapid rise in worldwide emissions of GHGs, the watering down of obligations under the Kyoto Protocol, and the emergence of scientific evidence demonstrating a need to stabilize concentrations as low as 350ppm are sufficient to raise serous doubts about the evolutionary potential of the protocol. There is a lot to be said for the proposition that what is needed at this juncture is a new approach to the regulation of greenhouse gases more in line with the real proportions of the problem (Victor 2001; Victor, House, and Joy 2005).

To avoid excessive pessimism, it is worth noting that there are signs of movement in this direction. In 2006, the EU heads of state made a public commitment to reduce GHG emissions by 20 percent relative to 1990 levels by 2020. The declaration issued at the end of the 2007 G8 meeting calls for initiatives designed to achieve deep cuts in emissions by 2050; the 2009 G8 meeting reaffirmed this call. Regional initiatives (e.g., California's 2006 legislation calling for a 20 percent cut in emissions by 2020) are emerging in many areas. Influential reports, such as the 2006 Stern Review, are producing calculations suggesting that the long-term damages expected to arise from climate change will exceed the costs of addressing the problem now by a wide margin (Stern 2007). There is a growing awareness that climate change could have major consequences, even at the level of high politics or national security.

Are these developments sufficient to license the conclusion that climate change has now reached a high enough level on the world's policy agenda to energize serious efforts to address the problem? If so, will this result in real innovations regarding the arrangements needed to confront the problem in a meaningful fashion, including a more substantive successor to the Kyoto Protocol? We cannot answer these questions in a decisive manner at this juncture. But whatever happens in the future, there is no avoiding the conclusion that the climate regime, at least over the period from the signing of the UNFCCC in 1992 to COP 15 at the end of 2009, fits the pattern I have called arrested development.

Analysis: Sources of Arrested Development

How can we account for arrested development in the case of the climate regime? Many observers take the view that the answer to this question lies in identifying factors that make climate change what some analysts characterize as a malign or wicked problem (Miles et al. 2002). The sources of GHG emissions are so deeply embedded in industrial societies and the projected costs of addressing the problem so large that it is hard to imagine countries devising an effective climate regime. But this approach is somewhat simplistic. By most estimates, the cost of controlling greenhouse gas emissions is actually well within our means; this cost would amount to only a fraction of what we routinely pay for national security. A coalition of the EU, the United States, and China would probably suffice to serve as an effective k-group, or set of first movers, regarding this issue (Schelling 1978). There are good reasons to believe that climate change has now risen to the level of high politics in the thinking of policy elites in many quarters.

An alternative argument, which I develop in the remainder of this chapter, focuses on the alignment of endogenous and exogenous factors. An ensemble of internal and external factors hobbles the regime, producing a poor fit between the biophysical and socioeconomic setting and the content of the regime. If I am right, we should think hard about the prospects for a major restructuring of the climate regime to produce a better fit rather than investing a lot of time and energy in making incremental adjustments to the arrangements established under the terms of the Kyoto Protocol.

Endogenous Factors

The UNFCCC and the Kyoto Protocol are not without virtues from a developmental perspective. Article 24 of the UNFCCC and Article 26 of the KP state simply, "No reservations may be made to the Convention" or to the protocol. As in the case of the ozone regime, there is also an effort to avoid the paralysis associated with reliance on unanimity in the formal sense as a decision rule. Article 15 of the UNFCCC and Article 20 of the KP allow the parties—as a last resort when all other efforts to reach agreement are exhausted—to adopt amendments by a three-fourths majority vote. The restriction on reservations is significant; many international agreements have been hobbled by reservations that member countries impose in order to build the coalition needed for ratification. The voting rule by contrast is less significant than it may appear to those of us schooled in democratic practices in which collective choices are made via majority voting. There is a pronounced tendency at the international level to proceed by way of consensus, whatever the formal provisions of the relevant agreement may say (Breitmeier, Young, and Zürn 2006). Because international regimes are heavily dependent on voluntary action on the part of member states when it comes to matters of implementation, it is important to avoid alienating key members via the use of decision-making procedures that seem coercive.

Two other features of the convention and protocol system are more problematic from the point of view of institutional development. Unlike the case of ozone, the climate regime does not include a provision allowing the COP or the MOP to make adjustments regarding matters already covered by the arrangement without triggering a need for ratification on the part of member states. Equally important are the handicaps arising from the provisions of Article 24 of the KP regarding entry into force. Whereas the UNFCCC entered into force following the fiftieth ratification, the KP entered into force only in 2005 after ratification by fifty-five states accounting for at least 55 percent of 1990 carbon dioxide emissions on the part of UNFCCC Annex 1 parties. Following the withdrawal of the United States in 2001, ratification on the part of Russia became a necessity for the KP to enter into force. Recognizing this fact, the Europeans made a number of concessions to entice Russia to join the regime. While this strategy ultimately worked from the perspective of entry into force, it led directly to unfortunate compromises reflected in the Marrakech Accords and subsequent efforts to move the KP from paper to practice.

The climate regime suffers as well from the fact that progress in fulfilling its goal is hard to measure. There is no way to operationalize the ultimate goal of avoiding dangerous anthropogenic interference in the Earth's climate system in a manner that is acceptable to all—or even most of—the parties. The Europeans have adopted the formula of preventing GHG concentrations from rising above 450ppm and capping temperature increases at 2°C. They deserve praise for their efforts to spell out the goal in concrete terms. But this approach leaves a lot to be desired. The two elements—GHG concentrations and temperature increases—may be incompatible. If the views of the IPCC are right, atmospheric concentrations of 450ppm may produce temperature increases of more than 2°C. The European position is arbitrary. It is an attempt to provide a quantifiable measure of progress that does not rest on a solid, much less generally accepted, foundation. Yet no one has a better approach to operationalizing the ultimate goal of the climate regime. Unlike the ozone regime, which seeks to prohibit the production and consumption of various chemicals and encompasses well-defined phaseout schedules, the climate regime lacks clear measures of success or progress. This limitation becomes particularly acute in the context of efforts to establish specific targets and timetables for reductions in emissions intended to mitigate the problem of climate change. In the absence of a more credible way to measure progress, the determination of targets for emissions reductions easily degenerates into an expediential exercise.

A factor that exacerbates this limitation arises from discursive differences regarding the framing of climate change as a policy problem. Is it better to approach the problem, as the Europeans tend to see it, as a matter of finding ways to decarbonize industrial societies or, as the Americans are apt to see it, as a matter of controlling levels of GHGs in the Earth's atmosphere? The two approaches have a good deal in common. But when it comes to efforts to implement the provisions of the regime on a day-to-day basis, important differences become apparent. Those seeking to control atmospheric concentrations are drawn to procedures featuring carbon sequestration, carbon capture and storage, and a variety of flexibility mechanisms; they tend to rely on technological solutions to solve problems of this sort. The goal of decarbonization, by contrast, leads to initiatives that force emitters, ranging from power plants to individuals and their modes of transportation, to think about feasible options for reducing their climate footprint and to make substantial changes in the character of industrial societies in the process. There is nothing inherently right or wrong with either of these approaches to the problem of controlling levels of GHGs in the Earth's atmosphere. But the two approaches can lead to substantial differences regarding the development of the climate regime. In the absence of some resolution of these differences, we can expect that the regime will become an arena for acrimonious debates about conflicting approaches to the problem or that critical players such as the United States will refuse to join the regime. In either case, efforts to move the regime onto the path of progressive development will falter.

While the climate regime, under the terms of the KP, has adopted a strategy generally referred to as targets and timetables, it is important to note that the protocol itself does not impose any requirements regarding the choice of policy instruments on the part of individual member states to meet their commitments. Once an Annex 1 country has accepted a specific target (e.g., Switzerland's target of 8 percent below 1990 levels by the end of the first commitment period), that country is free to pursue its goal within its own jurisdiction in whatever way it prefers. Any non–Annex 1 country that undertakes voluntarily to reduce its emissions is free to pursue its goals using policy instruments of its own choice. It would be incorrect, therefore, to describe the climate regime itself as a cap-and-trade arrangement in the ordinary sense of the term. From some points of view, this flexibility allowing individual members to decide for themselves how to meet their obligations under the KP is attractive. It recognizes both the diversity among member states with regard to their political and economic systems and the absence of any international mechanisms to enforce compliance on the part of member states. But this situation is also troublesome from the perspective of making measurable progress toward the reduction of GHG emissions, much less toward the overarching goal of solving the problem of climate change. Already, debates have emerged regarding the pros and cons of different approaches to the regulation of GHG emissions (e.g., the EU Emissions Trading Scheme versus the subnational approaches that have emerged in the United States). There is no way to engage at this stage in a rigorous evaluation of the relative merits of these arrangements with respect to their contribution to solving the problem of climate change. The use of a wide range of policy instruments in domestic settings is inevitable; it may even produce a range of useful natural experiments regarding alternative ways to restrict GHG emissions. But in the near future, we can expect to witness a chaotic process featuring tentative steps in a variety of directions that make it hard to evaluate the extent to which the climate regime is making progress toward protecting the Earth's climate system.

The KP itself does incorporate flexibility mechanisms adopted as a means of allowing the protocol to enter into force and spelled out in some detail in the provisions of the 2001 Marrakech Accords. Ironically, these mechanisms, including emissions trading, joint implementation, and the creation of the Clean Development Mechanism (CDM), were championed at an early stage by the United States but incorporated in various forms into the 2001 accords by others as part of a bargaining process focusing on the entry into force of the protocol following the withdrawal of the United States. There is a case to be made for the benefits of including these mechanisms. But they also have opened up opportunities for individual members or groups of members to pursue strategies detrimental to the cause of solving the problem of climate change. The stratagem of producing HFC-23 as a by-product of the production of HCFC-22 and then destroying it in a profitable manner under the provisions of the CDM is an illustrative case. So is the behavior of European firms that have raised prices to consumers of energy ostensibly in response to the establishment of the Emissions Trading Scheme, even though they received emissions permits free of charge in the initial allocation. My point here is not to denigrate the flexibility mechanisms that are allowed to operate under the terms of the KP and spelled out in more detail in the provisions of the Marrakech Accords. But at this stage, it is hard to avoid the conclusion that the flexibility mechanisms have generated confusion at best and opened up opportunities for manipulation at worst. The time may come when adaptive management will turn these mechanisms into assets from the perspective of problem solving. But for now, they add up to a factor that helps to explain why the climate regime fits the pattern of arrested development in contrast to progressive development.

Another endogenous factor that is relevant to the performance of the climate regime relates to funding. Both the UNFCCC and the KP include progressive language pertaining to funding. Article 11 of the UNFCCC calls for the establishment of a "mechanism for the provision of financial resources on a grant or concessional basis" and asserts that this "mechanism shall have an equitable representation of all Parties within a transparent system of governance." Article 11 of the KP goes on to spell out an obligation on the part of Annex 1 countries to provide "new and additional financial resources to meet the agreed full costs incurred by developing country Parties" in fulfilling their commitments under the climate regime and to limit increases in GHG emissions associated with economic development. But there is nothing in the climate regime to match the ozone regime's Multilateral Fund as an engine for progressive development. In practice, the climate regime has relied on the Global Environment Facility (GEF), a separate body managed jointly by the World Bank, UNEP, and the UN Development Programme (UNDP), to provide the funding needed by non–Annex 1 countries to devise and implement development strategies to meet their goals in a manner that minimizes the energy intensity of their production systems (Keohane and Levy 1996). There is no need to launch a general criticism of the GEF to see the limitations of this arrangement with regard to the problem of climate change. Not only is the GEF managed by officials who operate outside the culture of the climate regime but it is also subject to the politics arising from complex interactions among the World Bank, UNEP, and UNDP regarding the pros and cons of different development strategies and the proper procedures for administering and monitoring them. It is fair to observe that the funding needed to address the problem of climate change in an effective manner will exceed by an order of magnitude the funding needed to make a success of the MLF under the provisions of the ozone regime. Still, it is hard to avoid the conclusion that the nature and functionality of the current funding arrangements constitute an important endogenous factor in any effort to explain the performance of the climate regime (Keohane and Levy 1996).

The provisions of both the UNFCCC and the KP focus almost exclusively on what is known in the discourse on climate change as mitigation in contrast to adaptation. In some ways, this is a natural situation. There is something to be said for turning first to the prospects for mitigation, which directs attention to efforts to avoid the impacts of climate change, in contrast to adaptation, which is a matter of minimizing or alleviating the impacts of climate change once it occurs. Yet there is more to this story than meets the eye at the outset. Since developing countries and especially small island developing states are likely to be the major victims of climate change in its early stages, it is comparatively easy for decision makers in advanced industrial countries to look the other way when it comes to adaptation. All policy makers engage in a certain amount of wishful thinking, hoping that the projected impacts of climate change simply will not materialize, at least during their time in office. But this is a serious problem, given that both biophysical and socioeconomic impacts of climate change are already evident in some parts of the world (Arctic Council 2004). As a framework agreement, it is understandable that the UNFCCC does not include substantive provisions pertaining to adaptation. But the convention does not function as a catalyst encouraging key players to take an active interest in adaptation. The KP also has little to offer to those concerned with adaptation. Adaptation has surfaced now in COP and MOP deliberations, and there is a growing interest in developing a new protocol dealing with adaptation as a component of a post-Kyoto regime, a restructured climate regime to take effect when the KP's commitments expire at the end of 2012. Development along these lines is a possible trajectory for the climate regime. Still, it is hard to avoid the conclusion that the clear separation between mitigation and adaptation looms as an obstacle to be overcome by those interested in and desiring to set the climate regime on a path toward progressive development.

This raises the question of what will happen following the expiration of the KP's commitments at the end of 2012. In Article 3(9), for example, the protocol addresses commitments for "subsequent periods" and calls on the MOP to "initiate the consideration of such commitments at least seven years before the end of the first commitment period. . . ." The preferred method of adjustment is through revisions in Annex B on quantified reduction commitments on the part of countries that are signatories to the protocol. Beginning with presentations at the 2005 COP and MOP, a high-level consideration of this subject has emerged. It has become a top priority among the members of the regime as we move toward the end of the first commitment period. From the point of view of progressive development, this is good news. Yet the trajectory of this process—much less the conclusions likely to arise from it—is far from clear. There is a widespread sense that the overall politics of climate change are shifting significantly at this stage. The 2007 declaration of the G8 takes a relatively well-defined stand on the need for deep cuts in GHGs emissions (G8 Summit 2007). Major groups within the United States (e.g., the backers of the Western Climate Initiative and the Regional Greenhouse Gas Initiative among states in the northeast) have started to take serious steps to address the problem. Even the federal government in the United States has begun to show signs of moving the issue to a higher place on the policy agenda. The Bali Roadmap adopted at the end of COP 13 in December 2007 lays out a pathway intended to produce agreement on a post-Kyoto arrangement at COP 15 in December 2009. But the central question in this discussion of endogenous forces has to do with the extent to which the regime itself encourages the emergence of institutional changes that are progressive in nature. The record here is less encouraging. The relevant provisions of the KP clearly envision a process of incremental change in which targets and timetables are ratcheted up over time and some progress is made regarding issues like technology transfer, compliance, and funding. Yet the KP's Articles 20 and 21 make it clear that amendments will enter into force only for those regime members that provide written consent. The implication of this is that individual members of the regime can opt out of adjustments in targets and timetables or new policies and measures simply by failing to provide written consent regarding their acceptance of these changes. They even may be able to opt out of commitments that become inconvenient with the passage of time.

In some situations, a developmental process of this sort may be all that is needed. Much of the action in the ozone regime, for example, consists of adopting and revising phaseout schedules for specific ODSs. The issue of institutional adaptation and development with regard to climate is more fundamental in nature. As the evidence regarding both current and projected impacts of climate change piles up and is incorporated into the scientific consensus through the ongoing work of the IPCC, it is becoming apparent that addressing this problem effectively will require institutional adjustments that are more than incremental in nature. We can say now with confidence that restricting atmospheric concentrations of greenhouse gases to 450ppm will require very deep cuts in emissions over the next two to four decades (IPCC 2007); cutting back to 350ppm would require a dramatic shift from business as usual. It is apparent also that climate change already has begun to affect ecosystems and social systems in some places; there is no escaping the conclusion that these effects are destined to spread in spatial terms and to increase in intensity over time. The Earth's climate system is so complex that we cannot forecast the future trajectory of climate change and the impacts associated with it with any precision. Still, these observations suggest we need to make a sustainability transition with regard to institutional arrangements as well as economic and social practices. A future climate regime must not only mandate deep cuts in GHG emissions but also provide procedures that allow for responsible decision making under uncertainty. The existing provisions of the UNFCCC and the KP fall substantially short in these terms.

To conclude this account of endogenous factors in the case of the climate regime, I turn again to the question of leadership on the part of individuals. It is easy to find evidence regarding the role of leadership at various stages in the development of this regime. Examples that come to mind include the roles that Angela Merkel, then environment minister in Germany, played in hammering out the terms of the 1995 Berlin Mandate and that Raul Estrada, a diplomat from Argentina, played in the process that produced the Kyoto Protocol in December 1997. It is notable as well that the heads of the UNFCCC Secretariat—Michael Zammit Cutajar, Joke Waller-Hunter, and Ivo de Boer—have all been energetic and articulate actors able to make a difference in moving the climate regime from paper to practice, despite extreme limitations with regard to both material and political resources. We need to avoid exaggerating the importance of the contributions that these individuals have made, though. All of them have functioned largely as entrepreneurial leaders (Young 1991). There is little evidence that they have been able to stimulate the emergence of a new discourse in terms of which to reorient our thinking about the problem of climate change and to provide a new paradigm regarding the institutional issues associated with this problem. Even in entrepreneurial terms, their ability to affect the development of the regime has been limited. Estrada, for example, was limited severely by the difficulties of dealing with the United States regarding the terms of the Kyoto Protocol. Even so, we should not ignore the contributions of these and other individuals to the development of the climate regime; they turned in outstanding performances in situations that often were not conducive to progressive development. Looking to the future, it is encouraging that Ban Ki-moon, the current UN Secretary General, has identified climate change as a top priority for his period in office.

Exogenous Factors

To balance the preceding account, I turn now to exogenous factors that have interacted with the endogenous factors to generate the pattern of arrested development that characterizes the climate regime. Many of those who seek to understand the problem of climate change come away with the sense that this is the most malign or wicked environmental problem facing the world today and that designing and implementing an international regime capable of dealing with the problem effectively is beyond our capacity to engage in problem solving on a global scale (Miles et al. 2002). There is some basis for this interpretation. Reliance on fossil fuels is pervasive and deeply embedded in industrial societies. A solution to the problem of climate change would amount to the production of a public good, so that we must expect to encounter free-ridership on a large scale in this realm. Some powerful players (e.g., the multinational oil companies) resist efforts to address the problem on the grounds that they expect to be losers in any process that lowers the dependence of modern societies on fossil fuels. The probable distribution of disruptive impacts associated with climate change is such that costs will fall heavily on victims who have not caused the problem and have comparatively limited capacity to exercise influence on a global scale in contrast to those who have caused the problem but have less to worry about regarding its consequences.

Still, there are good reasons to question the proposition that the problem of climate change is simply too hot to handle or too malign to solve. Much of the technology needed to wean societies off a heavy dependence on fossil fuels already exists. Experience with other issues (e.g., ozone depletion and acid rain) suggests that the corporate world is perfectly capable of coming up with major technical advances in relatively short order once business leaders become convinced that the decision has been made to phase out the use of ODSs, to cut sulfur emissions drastically, or to deal with other well-defined issues. There is no reason to believe that the case of GHG emissions will turn out to be fundamentally different in these terms. Some leading corporations (e.g., DuPont, 3M, and GE) already are investing heavily in products expected to cut greenhouse gas emissions substantially; the establishment of the United States Climate Action Partnership (USCAP) is significant in these terms. What business leaders need to trigger a swing into high gear in addressing the problem of climate change is a clear understanding that society is determined to take steps to address the problem, combined with reasonable assurance that the playing field will be level in the sense that their competitors will confront the same regulatory setting.

What this means is that the vital core of the problem of climate change centers on the development of sufficient political will to tackle the problem effectively. It is worth reminding ourselves that there are precedents for such actions, including addressing some problems that are or were large scale and transboundary in scope. The Marshall Plan of the late 1940s and early 1950s, which involved transfer payments on a massive scale from the United States to Europe and required bipartisan political support in the United States, is a case in point (Mills 2008). A comparable investment (roughly 1% of GDP) in today's dollars would go a long way toward alleviating the problem of climate change. Estimates by the well informed make it clear that taking steps to come to terms with climate change would cost only a fraction of what the United States routinely allocates to defense spending. There is little prospect of devising an effective method of coming to terms with climate change without the full and active engagement of the United States and China, which together account for about 35 percent of all GHG emissions. But circumstances may converge before long that will make a coalition of China, the EU, and the United States regarding the problem of climate change a real possibility.

However we approach the issue, progress in this realm will depend on finding ways to deal with a variety of economic, cognitive, political, and institutional obstacles that are external to the climate regime but likely to play critical roles as determinants of the effectiveness of this regime. The technological and financial resources needed to solve the problem of climate change already exist. But entrenched institutional arrangements impede efforts to make progress in this realm. One obstacle arises from a system of rules that does not require full-cost accounting regarding emissions of GHGs. The use of the atmosphere as a repository for wastes or residuals is for all practical purposes a free factor of production. Whereas other industries take it for granted that they must pay for the disposal of wastes through the use of landfills, incineration, or other methods of disposal, producers of GHG emissions are allowed to use the atmosphere as a repository for wastes at no cost to themselves. Faced with a situation in which one factor of production is free while others are costly, the rational choice is to use as much of the free factor as possible. The solution to this problem is to change the rules of the game to make users pay for the use of this ecosystem service. Such a change would drive up market prices of some goods and services (e.g., electricity and gasoline) as producers seek to pass on costs from themselves to consumers. But there is nothing inherently wrong with this outcome.

Three economic issues are worth noting in this context. It is much easier to compute the costs of taking steps to avoid climate change than to calculate the benefits attributable to taking these steps. Many of the benefits (e.g., the avoidance of various health problems) will accrue over long periods, a fact that makes it critical to make use of appropriate discount rates in thinking about such matters. As the vociferous debate over the use of discount rates in the Stern Review makes clear, there is room for extensive debate among economists (and others) regarding this matter (Toll and Yohe 2006). A second factor concerns the incidence of the benefits and costs associated with efforts to address the problem of climate change. Most analysts agree that the impacts of climate change—at least in the short run—will be most pronounced in developing countries, whereas the lion's share of the costs of dealing with climate change must be assumed by advanced industrial countries, both directly in the form of adjustments to their own economic systems and indirectly in the form of aid to developing countries willing to regulate their GHG emissions. One result of this will be a need to make transfer payments to non–

Annex 1 countries on a sizable scale. Agreeing on the nature and scope of transfer payments is hard enough in domestic settings; we have limited experience regarding the treatment of such matters on a global scale. Finally, the fact that a solution to the problem of climate change would amount to the production of a public good poses a problem that we cannot ignore. Many societies do find ways to supply public goods on a large scale, despite the influence of the logic of collective action leading to the familiar problem of free-ridership (Olson 1965). In the case of climate change, it seems likely that the most realistic solution to this difficulty would involve forming a k-group of major players with incentives to provide leadership regarding this issue (Schelling 1978). A coalition of the United States, the EU, and China might well be sufficient to play this role effectively. The addition of Brazil and India would surely make the coalition sufficient for this purpose.

Given these economic challenges, cognitive factors exogenous to the regime itself are likely to play a critical role in efforts to solve the problem of climate change. What is needed is a reshuffling of the conceptual landscape that yields a new paradigm to structure our thinking regarding climate change. This is a tall order. The past ten thousand years have been exceptionally benign with regard to the behavior of the Earth's climate system. No one living today has experience with extreme volatility in the behavior of this system. It is hard for ordinary people to grasp the true scope of the problem of climate change. Not only are most people preoccupied with day-to-day challenges like earning a living, procuring adequate health care, and providing for the education of their children but they also lack the knowledge required to understand the problem of climate change in any more than a rudimentary manner. Then, too, there are the profound uncertainties surrounding the issue of climate change. How seriously should we take the prospect of rapid climate change events (RCCEs) (Mayewski and White 2002)? Is there a danger that emphasizing the prospect of abrupt change as a means of galvanizing public opinion will backfire when such changes fail to materialize in the near future? The common thread running through these concerns is the absence of a clear-cut and easily understandable paradigm that encourages people to think of the Earth and its climate system—including the role of human actions—as a large, complex, and dynamic system. It is impossible to predict if and when such a paradigm shift will occur, much less what form a new paradigm would take. Even so, there are reasons to believe that such a transition will occur in the foreseeable future. The reports of the IPCC and related groups (e.g., the Arctic Council's Arctic

Climate Impact Assessment) are now providing incontrovertible evidence that we are experiencing the onset of climate change and that the impacts are likely to be nasty from a human perspective. Both the existence and the importance of threats to the Earth's climate system are coming into focus in the minds of opinion leaders who are not specialists in the science of climate change. Climate change is emerging as a significant issue in electoral politics. None of this makes a paradigm shift inevitable. But it does make such a shift regarding climate change a distinct possibility.

This brings us back to the politics of climate change. A number of political factors constitute impediments to making real progress in this realm. At least in democratic systems, timing is a major consideration. Electoral politics focus on immediate concerns (e.g., jobs, health care, or social security) that affect the lives of voters today. So long as climate change is perceived as a distant problem, it is easy for elected officials to set it aside as a not-on-my-watch issue. This is why exogenous shocks, like the deaths of elderly people associated with the European heat wave in 2003, can make a difference in the emergence of such issues on the public policy agenda. The behavior of the United States and the standoff between the United States and leading developing countries like China are also major political factors affecting responses to climate change. Despite the fact that together they account for roughly 35 percent of all GHG emissions, the United States and China continue to engage in a dialog of the deaf regarding this issue. China is unwilling to make serious commitments regarding climate change so long as the United States has not accepted obligations regarding this issue. The United States takes the view that it should not be expected to make real commitments so long as the non–Annex 1 countries and especially the large ones like China are not prepared to make commitments regarding climate change. The result is gridlock in policy terms.

Beyond this, issues of fairness or equity loom large in the politics of climate change. A prominent example involves initial allocations of permits arising in conjunction with the popular cap-and-trade approach to controlling emissions (Raymond 2003). So long as emissions permits are grandfathered—allocated to existing emitters free of charge—it is hard to see any effective way forward in coming to terms with the problem of climate change. If we think of the atmosphere as part of the common heritage of humankind, giving away its ecosystem services to polluters free of charge is simply unacceptable. Such a procedure cannot gain the support of those in the developing world who are likely to interpret such a giveaway as a form of ecoimperialism. Even at the domestic or regional level, the political repercussions are serious. Some European firms that received allocations of emissions permits free of charge during the initial stage of the EU's Emission Trading Scheme (EU ETS) raised prices to their customers and claimed it was justified due to higher costs of production caused by the establishment and operation of the EU ETS (Kruger and Pizer 2004). The EU ETS has now been reformed, at least in part to deal with concerns about the equity of the system. But the general point is clear. So long as issues of this sort remain unresolved, it is unlikely we will be able to cross the threshold leading to sustainability with regard to climate change. This is where the issue of a paradigm shift becomes critical. It is hard to see an effective way to deal with these issues of equity and fairness in the context of business-as-usual practices. But a new paradigm emphasizing the nature of the climate system as a large, complex, and dynamic system whose behavior will affect the welfare of people everywhere could open the way toward reformulating climate change as a top priority on the global policy agenda (Litfin 1994).

We need to understand as well that the climate regime is not a self-contained arrangement that has little or no interaction with other institutional arrangements operating at the international level. Interactions between the ozone regime and the climate regime are particularly striking. As Velders and colleagues observe, "reductions in atmospheric ODS concentrations, achieved to protect ozone, also serve to protect the climate" (Velders et al. 2007: 4814). Detailed calculations indicate that accelerating the phaseout of various ODSs under the terms of the ozone regime will result in reductions of 135 billion tons of CO_2 equivalent over the period between 1990 and 2010. These reductions exceed those expected to accrue from all reductions mandated under the terms of the Kyoto Protocol through the end of the first commitment period. This is a remarkable example of what observers have characterized as positive or synergistic interplay between regimes that have no formal links to one another (Oberthür and Gehring 2006).

Still, interactions between these regimes also can generate perverse incentives that are detrimental to efforts to address the problem of climate change (Kaniaru et al. 2007). A striking example concerns the production and destruction of HFC-23. A powerful greenhouse gas, HFC-23 is a natural by-product of the production of HCFC-22, which is a permissible short-term substitute for some ODSs phased out under the terms of the Montreal Protocol (Parson 2003). Until recently, because HCFCs were not scheduled for phaseout until 2030 in developed countries and 2040 in developing countries, firms seeking to comply with the requirements of cap-and-trade arrangements, such as the EU ETS, were able to gain credit in the form of carbon offsets by paying for the destruction of HFC-23 under the auspices of the climate regime's Clean Development Mechanism. What made the situation particularly perverse was that the value of the offsets accruing to individual firms was much larger than the cost of producing and then destroying HFC-23 as a by-product of HCFC-22. As a result, developing countries—especially China—had an incentive to produce HFC-23 simply in order to reap gains by selling the destruction of this by-product to firms operating under the auspices of the CDM (Kaniaru et al. 2007). The agreement reached in September 2007 by the MOP of the Montreal Protocol to take action to curb this practice was therefore an important breakthrough. This type of unintended side effect involving interactions between the ozone and climate regimes is illustrative of a type of institutional interplay likely to occur with increasing frequency as we move toward a system featuring a growing density of regimes operating at the international level. Efforts to solve these problems also illustrate the challenges involved in supplying international governance in a world that lacks an overarching governance system possessing the authority to iron out such problems of institutional interplay as they arise in specific issue areas.

Nor is this the only form of interplay likely to arise as we move farther down the track of implementing the Kyoto Protocol and constructing a more effective successor to take effect following the end of the first commitment period. A particularly important concern involves trade-environment interactions. We can foresee a range of situations in which companies move certain operations offshore to escape tighter restrictions on emissions in Annex 1 countries, emissions trading schemes lead to poorly regulated forms of international trade, proposals aimed at restricting trade in products featuring carbon-intensive production processes gain traction on policy agendas, or countries claim exemptions from general WTO rules on grounds relating to efforts to address the problem of climate change. Similar observations are in order regarding interactions with the regime for biological diversity relating, for instance, to the development and implementation under the auspices of the climate regime of arrangements dealing with the role of forests in sequestering carbon dioxide.

None of this is to say that problems arising from institutional interplay are unavoidable in efforts to address the problem of climate change or that those responsible for administering the regimes in question will be unable to find effective ways to resolve such problems when they do arise. But it is important to recognize institutional interplay as an exogenous factor that will need to be addressed to ensure success in efforts to address climate regime.

Endogenous-Exogenous Alignment

Climate change is an enormous problem, so much so that some analysts have concluded that international society does not have the capacity to supply effective governance. I argue in earlier sections of this chapter that this proposition is not convincing. The technological and financial resources needed to come to grips with climate change already exist. The cost would be considerably less than what we invest routinely and with little debate to produce the public good known as national security or defense. Numerous technologies that could help to reduce GHG emissions are either available already or well within the capacity of applied research to bring on stream. A number of the steps needed to address the problem of climate change might well have positive cost-benefit ratios and thus add to the GDPs of participating countries.

When it does come time to make a concerted effort to break the prevailing pattern of arrested development regarding the climate regime, how should we proceed? My answer to this question focuses on improving the alignment of the endogenous and exogenous factors discussed in the preceding subsections to develop a climate regime that is well matched to the biophysical and socioeconomic settings in which it operates. What we face today is a problem featuring nonlinear and often unpredictable changes coupled with a regime that is sluggish and lacking in the nimbleness needed to address these changes. There is much to be done in this connection that is incremental in the sense that it does not require scrapping or drastically altering the existing regime. But in the end, more far-reaching changes affecting exogenous as well as endogenous matters are likely to be needed to break out of the existing pattern. Faced with this situation, we can and should pursue incremental improvements while preparing for more fundamental changes to be made whenever a window of political opportunity opens in this realm.

Incremental changes that would improve the match between endogenous and exogenous factors in this case fall into two general categories: (1) adjusting existing arrangements to make them more nimble, or better able to cope with the dynamics of the Earth's climate system, and (2) adding components designed to address matters of adaptation.

The Earth's climate system is highly dynamic. We must expect it to change in ways that are nonlinear, sensitive to initial conditions, irreversible, and from time to time abrupt. The climate regime, on the other hand, is relatively rigid; it responds sluggishly to new inputs regarding the behavior of the climate system. This feature of the regime reflects the determination of member states to pay more attention to protecting their sovereign rights—as articulated in places like Principle #2 of the 1992 Rio Declaration—than to putting in place a governance system that is nimble enough to respond quickly and effectively to occurrences unforeseen by the regime's creators. A useful initiative in this connection would be to create a more sophisticated monitoring system capable of providing early warning of impending changes in the climate system. But early warning by itself will be of little help in the absence of a capacity to respond promptly and vigorously in institutional terms to evidence that major changes in the climate system are occurring already or are imminent. There may be something to learn in this realm from the experience of the ozone regime in which decisions regarding phaseout schedules for chemicals already regulated can be taken more quickly and easily than decisions about adding new chemicals or families of chemicals to the list of controlled substances.

No matter how much progress we make in these terms, however, we still will need to make decisions about climate-related matters under conditions of uncertainty (Tversky and Kahneman 1974; Kahneman 2003). In high-stakes situations, such as climate, it seems desirable to avoid settling for a decision-making procedure that ignores low-probability events or dismisses them as a consequence of adopting conventional approaches to discounting. It does not pay to fixate on future events that may or may not occur, especially when current pressing concerns require attention. But climate change exemplifies a category of problems with regard to which it is prudent to invest resources now as a means of minimizing future costs or avoiding them altogether. A similar rationale underlies defense expenditures, which are based to a large extent on the logic of deterrence and which exceed by a sizable margin the likely costs of taking action to avoid dangerous anthropogenic interference in the Earth's climate system. Thoughtful observers disagree about how to conceptualize, much less how to compute, the probable costs of climate change. Yet responsible estimates suggest that these costs could easily exceed the costs of mitigation by an order of magnitude (Stern 2007). Responding to uncertainty in a precautionary fashion makes sense in a situation of this sort.

We could make these adjustments to improve the alignment of endogenous and exogenous factors without changing the basic architecture of the UNFCCC and the KP. When it comes to adaptation, by contrast, we can expect to hear increasingly strong and persistent calls for the addition of a new component to the existing regime focused on adaptation. Some parts of the world already are experiencing serious impacts of climate change; these impacts will spread and intensify during the coming years. Yet the existing regime focuses largely on mitigation—efforts to avoid the occurrence of climate change by reducing emissions of GHGs rather than to prepare for or respond to the impacts of climate change. There are good reasons to continue to emphasize mitigation. But it is now time to address the issue of adaptation in a focused manner. What could or should an adaptation strategy encompass? The answer to this question varies greatly from place to place. But some broad concerns that deserve attention in this realm are already emerging. These include reinforcing or rebuilding infrastructure (e.g., roads, airports, and coastal barriers) disrupted by climate change, preparing for climate-related extreme events (e.g., tropical storms and floods), addressing issues of human health associated with climate change (e.g., the death of elderly shut-ins resulting from extreme heat), combating the impacts of climate change on ecosystems (e.g., the destruction of spruce forests in Alaska), and developing new insurance schemes to protect the equity of those likely to be affected by climate change (e.g., finding ways to insure homes against intensified coastal storm surges or wildfires). In the final analysis, many of these issues will require attention on the part of individual members of the climate regime and even regional or local authorities within individual members.

What would be the value added from establishing broader—international or global—arrangements dealing with adaptation? The answer to this question is not yet entirely clear. But it could involve such things as serving as a clearing house to promote exchanges of information regarding effective responses to the impacts of climate change, providing financial assistance to those ready to address the issue of impacts explicitly, and encouraging large-scale (re)insurance companies (e.g., AIG, Zurich Financial Services, Munich Re, or Swiss Re) to experiment with new forms of insurance available to those most likely to be affected by climate change.

It is encouraging to see the growing interest in adaptation among those who are working to hammer out agreement on the terms of a successor to the KP to take effect following the end of the first commitment period. There is a long way to go in coming to terms with adaptation. Annex 1 countries have a tendency to interpret a focus on adaptation as a lack of commitment to mitigation on the part of developing countries; developing countries sometimes see a focus on adaptation as a way of diverting attention from the need to reduce GHG emissions on the part of key Annex 1 countries. But adaptation is now firmly established as a priority issue on the climate agenda. We can expect a serious effort to devise the terms of an adaptation protocol for inclusion in the next phase of efforts to flesh out and strengthen the climate regime.

Suppose now that the onset of climate change continues or even accelerates and that an increase in climate-related disasters opens up opportunities to reform the climate regime in more substantial ways or even to restructure its basic premises. What changes should we recommend to improve the alignment of internal and external factors affecting the performance of this regime? The answer to this question includes several distinct elements. It would help to start by requiring all parties concerned to make a firm commitment to several major principles, including full-cost accounting for ecosystem services provided by the atmosphere, common but differentiated obligations on the part of member states, and a precautionary (if not a worst-case) approach to uncertainty regarding the behavior of the Earth's climate system. A binding commitment to the achievement of deep cuts in emissions, of the order of 60 to 80 percent relative to 1990 levels by 2050, should accompany the adoption of these principles. As in the case of ozone, it would be helpful to develop a mechanism for accelerating reduction schedules without triggering a need for formal ratification on the part of individual member states. This would require, among other things, the establishment of a climate fund analogous to the MLF in the ozone regime; it is not at all clear whether relying on the GEF to administer this fund would be a good idea.

A restructured climate regime should avoid becoming too prescriptive regarding choices of policy instruments individual member states make to fulfill their international obligations. It is not obvious that a cap-and-trade system operating on a global scale would constitute a preferred arrangement; some members might opt for cap-and-trade systems at the domestic level, while others opt to use different procedures to bring down emissions. A restructured regime would need to encompass flexibility mechanisms, such as the current arrangements known as emissions trading, joint implementation, and the Clean Development Mechanism. As experience under many other regimes makes clear, however, such mechanisms are often subject to manipulation and may produce unintended and undesirable results (e.g., the HFC-23 problem in conjunction with the CDM). Differences among member countries regarding both political and economic systems and policy cultures are so large that there is no alternative to giving them the option of making use of policies and measures of their own choosing in this setting. The issue of compliance, on the other hand, does require attention at the overall regime level. The parties to the KP are already taking significant steps under the terms of Article 18 to address the problem of noncompliance. These include the idea of imposing penalties during the next commitment period on those who fail to meet their obligations during the current commitment period. It is hard to see how this procedure would work in practice, however, especially as applied to larger players in the system (e.g., Japan, China, and the United States). In devising a restructured regime, there will be a clear need for upgrading provisions regarding compliance, drawing on what some analysts describe as the management approach as well as the enforcement approach to eliciting compliance at the international level (Chayes and Chayes 1995).

Some of these suggestions are likely to seem unrealistic or even utopian viewed from the perspective of the existing climate regime. But there is considerable evidence to suggest that policies—including those dealing with environmental regimes—exhibit a pattern of punctuated equilibrium in the sense that extended periods of stasis are interrupted occasionally by more or less far-reaching and dramatic changes (Baumgartner and Jones 1993; Repetto 2006). There are some indications that we are approaching a tipping point with regard to the problem of climate change that will open up such an opportunity for major institutional changes. The case for being prepared to seize such an opportunity to improve the alignment of internal and external factors affecting the performance of this regime is compelling.

Forecast: The Road Ahead

There is no way to avoid the conclusion that the climate regime, at least at this stage in its development, fits the pattern I have labeled arrested development. As figure 4.1 shows, GHG emissions for Annex 1 countries other than those with economies in transition rose 9.9 percent between 1990 and 2006. Canada's emissions in 2006 were 21.7 percent above the 1990 level, and Japan's emissions were 5.3 percent above the 1990 baseline. Among the Annex 1 countries that have ratified the KP, only the members of the EU-15 have a realistic prospect of meeting their KP commitment of 8 percent reductions in 1990 emissions levels during the first commitment period.

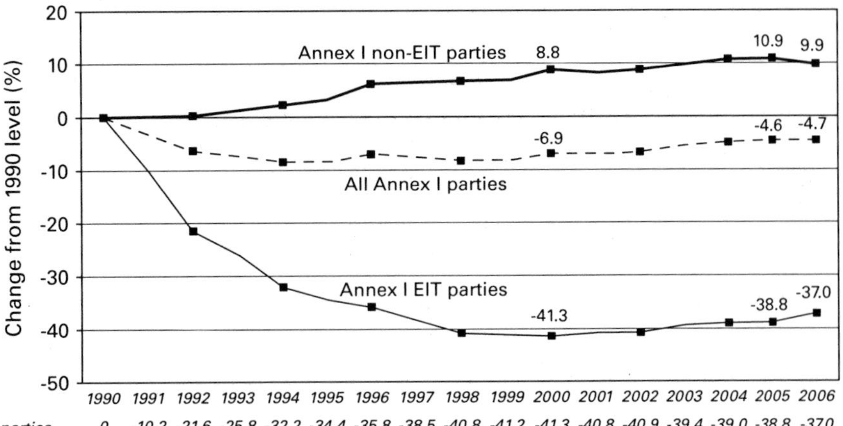

	1990	1991	1992	1993	1994	1995	1996	1997	1998	1999	2000	2001	2002	2003	2004	2005	2006
Annex I EIT parties	0	-10.2	-21.6	-25.8	-32.2	-34.4	-35.8	-38.5	-40.8	-41.2	-41.3	-40.8	-40.9	-39.4	-39.0	-38.8	-37.0
Annex I non-EIT parties	0	-0.1	0.3	1.0	2.3	3.3	6.2	6.3	6.6	6.7	8.8	8.1	8.7	9.9	10.7	10.9	9.9
All Annex I Parties	0	-3.2	-6.6	-7.4	-8.4	-8.4	-6.9	-7.7	-8.2	-8.3	-6.9	-7.2	-6.8	-5.5	-4.9	-4.6	-4.7

Figure 4.1
Greenhouse gas emissions excluding land use, land-use change, and forestry. *Note*: EIT = economies in transition. *Source*: United Nations Framework Convention on Climate Change Web site, http://unfccc.int/ghg_data_unfccc/items/4146.php.

Treated as a group, the EU-15 had made actual cuts in emissions by 2007 of around 3 percent relative to 1990 levels. Attributable in part to special circumstances involving the reunification of Germany and the end of coal mining in the UK during the 1990s, this overall performance includes increases in emissions on the part of a number of EU member states. At this stage, "European Environment Agency projections suggest that current policies will leave this picture unchanged by 2010" (UNDP 2007: 54).[2] To meet its KP target by the end of the first commitment period, the EU-15 will need to rely on a combination of sequestration through changes in land use practices and offsets, including credits for projects carried out under the auspices of the Clean Development Mechanism (European Environment Agency 2008).

The picture is bleaker with respect to the United States, which has not accepted any international obligation to reduce its GHG emissions, and especially a number of countries that are members of the climate regime but are not included in Annex 1 of the UNFCCC. Although U.S. emissions have leveled off and are not expected to grow significantly in the future, they still reached a level 14.4 percent above the 1990 baseline by 2006. China's emissions in 2004 were 47 percent above the 1990 level, and India's emissions at the same time had increased to 55 percent above the 1990 baseline (Wikipedia 2007; European Environment

Agency 2008). Currently, Chinese emissions are thought to be rising at a rate of about 4 percent per year, though this figure is subject to considerable variation from one year to the next (Congressional Research Service 2008). There is general agreement that China has now moved into first place as the world's leading emitter of GHGs.

How rapidly would emissions of GHGs have risen in the absence of the UNFCCC and especially the KP? There is no way to test institutional counterfactuals of this sort rigorously (Tetlock and Belkin 1996). Some members of the regime have made a good-faith effort to take their obligations seriously, devising considered policies to move toward compliance with the reduction targets assigned to them in Annex B of the protocol. It is possible that emissions would have risen farther and faster in the absence of agreement on the KP's targets and timetables. We cannot move beyond informed judgment in this context. What we can say with confidence is that the regime has not put us on track to meet the modest goals of the KP, much less to make headway in solving the problem of climate change more generally. Given the findings set forth in the IPCC's fourth assessment report indicating the need for reductions in emissions of the order of 60 to 80 percent to meet the UNFCCC's goal of avoiding dangerous anthropogenic interference in the Earth's climate system and the emerging view within the scientific community that we should be thinking in terms of stabilizing at 350ppm rather than 450, it is hard to avoid the conclusion that arrested development is the prevailing pattern in this case (IPCC 2007).

Those searching for a more optimistic conclusion regarding the regime's track record generally treat it as a first step or a work in progress and take heart from the idea that the regime has started the ball rolling toward the development of a more substantial regime during the foreseeable future. This incrementalist perspective is based implicitly—if not explicitly—on the premise that environmental regimes have a built-in dynamic producing step-by-step development from modest beginnings to the emergence of more extensive and effective arrangements over time. The ozone regime provides a paradigmatic example of this way of thinking. But how persuasive is this argument when applied to the climate regime? In my judgment, this perspective is a hard sell in the case of climate change. In part, this judgment stems from the mismatch between a complex and dynamic climate system likely to produce surprises in the form of nonlinear, irreversible, and sometimes abrupt changes and a regime that is far from nimble in responding to such changes, much less anticipating them and responding appropriately in a timely manner.

Partly, it arises from accumulating scientific evidence indicating that deep cuts in emissions of the order of at least 60 to 80 percent below 1990 levels will be needed within the next couple decades if we are to fulfill the goal articulated in Article 2 of the UNFCCC of avoiding dangerous anthropogenic interference in the Earth's climate system.

Arguably, the most important contribution of the KP involves the experience it has stimulated with the design and administration of a number of incentive mechanisms (e.g., the EU ETS and the Chicago Climate Exchange [CCX]) developed to guide the behavior of emitters of GHGs. Any effective effort to make deep cuts in GHG emissions will need to provide a variety of actors—entrepreneurs, investors, established firms—with opportunities to gain from initiatives that lead to reductions of GHG emissions. Although there is much more to be learned about such mechanisms, the arrangements that have emerged in conjunction with the KP constitute a step in the right direction (Aldy and Stavins 2007). The operation of the European trading scheme in particular has revealed a number of significant issues that will require attention in creating effective markets or quasimarkets in permits as mechanisms for reducing GHG emissions. Many analysts are seeking to extract lessons from this experience and to devise improved incentive mechanisms that could be used in the next stage of the climate regime.

Even so, it is hard to explain away the fact that many of the UNFCCC Annex 1 countries that have ratified the KP and accepted binding obligations to cut emissions will fail to meet even the modest commitments spelled out in Annex B of the protocol, as modified by the Marrakech Accords. Those countries that are not listed in Annex 1 (e.g., China and India) or that have not ratified the protocol (e.g., the United States) have increased their emissions and, in several cases, are continuing to do so at a rapid pace. It would be naïve, given these circumstances, to expect the existing regime to develop in a stepwise fashion that resembles the pattern of the ozone regime and that would turn it into a case of progressive development.

The conclusion I draw from this discussion is that progress in the development of an effective climate regime will require a nonlinear or step-level change toward a successor to the existing regime in contrast to a more gradual evolution based on developments unfolding within the existing regime. To add a note of optimism, the available evidence suggests that such a change is coming slowly into focus as a distinct possibility. As scientific certainty regarding the reality of climate change and experience regarding the impacts of this change increase, members of the policy community are beginning to push the issue of climate change toward the top of the policy agenda. The 2006 Stern Review characterized climate change as the largest externality ever and argued that the likely costs of climate change far exceed the costs of taking steps to avoid it (Stern 2007). European heads of state have made a public commitment to a 20 percent reduction of GHG emissions by 2020; the declaration setting forth the conclusions of the 2007 G8 meeting calls for halving GHG emissions by 2050 (G8 Summit 2007), and the 2009 declaration goes even further, calling on developed countries to reduce "emissions of greenhouse gases in the aggregate by 80% or more by 2050 compared to 1990 or more recent years" (G8 Summit 2009: paragraph 65). Even the United States, with a new administration in place, has joined the call to reach agreement on a successor to the KP that has teeth.

Meanwhile, a growing number of subnational bodies—states, provinces, cities, and so forth—have moved to take the issue of climate change into their own hands and to initiate programs designed to lead to deep cuts in GHG emissions in a timely manner (Betsill and Rabe 2009). In the United States, where the federal government has been a laggard on this issue, particularly striking developments include legislation on the part of California and coordinated action on the part of states in the northeast, known as the Regional Greenhouse Gas Initiative (see www.rggi.org), and in the West, known as the Western Climate Initiative (see www.westernclimateinitiative.org). We cannot assume that these forces will coalesce to trigger fundamental change in the existing climate regime in the immediate future. But as those who think about these issues in terms of the idea of punctuated equilibrium argue, distinct forces do occasionally coalesce or align with one another in ways that allow for transformative changes in public policies and in associated institutional arrangements (Baumgartner and Jones 1993; Kingdon 1995; Repetto 2006). It is not far-fetched to expect that some such convergence could occur in the coming years as policy makers seek to reach agreement on new terms for the climate regime following the expiration of the KP's commitments at the end of 2012. We should at least be aware of this possibility and make an effort to determine what changes to recommend in the event that a window of opportunity for more far-reaching change does open in the next few years.

Conclusion

In this chapter I make the case that the climate regime fits the emergent pattern of institutional change I have labeled arrested development and that this is, for the most part, a consequence of misalignment between the nature of the problem and the character of the regime. This misfit is, first and foremost, a function of the fact that the relevant biophysical systems are dynamic and prone to changes that are nonlinear, irreversible, and surprising, whereas the regime is sluggish and unable to respond in a nimble fashion to such changes in the climate system. This problem is exacerbated by the growth of scientific knowledge indicating that the effects of climate change are likely to be severe and costly and that we do not have the luxury of considering the problem in a leisurely fashion, responding only after we are absolutely certain about the likely consequences of the problem. There is an emerging consensus within the scientific community as well as large segments of the policy community that we must act to tackle this problem within the coming decade or the next two to three decades at most in order to fulfill the UNFCCC's goal of avoiding dangerous anthropogenic interference in the Earth's climate system.

Does this mean we are doomed to failure and consequently to suffer the consequences of drastic climate change? We need to recognize this as a real possibility. The records of both prehistoric and historic civilizations demonstrate that great societies have failed to address major challenges relating to human-environment relations on a fairly regular basis and have suffered severe setbacks or even collapsed as a consequence of this failure (Tainter 1988; Diamond 2005; Homer-Dixon 2006). Despite our advanced economic, political, and technological systems, we should not take it for granted that the dramatic failures of the past are of no relevance to our own civilization. I do not conclude from these observations that we must resign ourselves to gloom and doom in thinking about the problem of climate change. What is needed are decisive initiatives aimed at strengthening the existing governance system covering emissions of greenhouse gases. Efforts to craft agreement on the terms of a successor to the Kyoto Protocol seem agonizingly slow and inappropriately modest. Yet there are signs that we are now moving toward a window of opportunity that would allow us to break the pattern of arrested development in this realm. To the extent that this is the case, it behooves us to prepare carefully crafted proposals for a more effective climate regime so that we can swing into action quickly and effectively when the time comes to make a quantum leap in the arrangements we rely on to come to terms with the problem of climate change.[3]

5

Diversion: The Regime for Whales and Whaling

Overview: The Big Picture

Indigenous peoples have hunted whales, or, to use current terminology, engaged in "aboriginal subsistence whaling" for millennia. Historically, these shore-based and small-scale practices did not have an appreciable impact on populations of targeted whales, like the bowhead whale (*Balaena mysticetus*) hunted by Alaskan and Russian natives mainly in the Beaufort and Chukchi seas. Today, this picture is cloudier. Subsistence harvesting combined with other threats to whales can make a difference in the status of specific cetacean stocks. The regime established under the International Convention for the Regulation of Whaling (ICRW) requires subsistence hunters to rely on traditional means of hunting and to refrain from selling whale products. But the gap between the ideal and the actual is significant on both scores. This is particularly true in the case of harvesting methods where groups concerned with humane methods of killing have pushed hard for the introduction of improvements, such as the use of penthrite grenades. Still, aboriginal subsistence whaling is alive and well in a number of places around the world (Caulfield 1997). Even the United States, which is otherwise a strong advocate of terminating whaling, has thrown its weight behind measures to approve whaling on the part of Alaska Natives in the deliberations of the International Whaling Commission (IWC) and other forums in which the condition of whale stocks is debated.

The impetus behind efforts to establish an international governance system to conserve whale stocks and to regulate the harvesting of whales was a response to the impacts of commercial or industrial whaling. Starting with Basque whalers as early as the eleventh century and carried out initially as a shore-based activity, industrial whaling grew steadily over the ensuing centuries. In the nineteenth century, whalers roamed the high seas, pursuing whales during voyages lasting up to three or four years (Dolin 2007). By the early twentieth century, whalers had developed the whaling cannon or harpoon gun and initiated a system of whaling involving clusters of catcher boats linked to large steam-powered factory ships capable of remaining on the whaling grounds for months or even years on end. As in other areas involving the human use of renewable resources, the growth of harvesting power coupled with the absence of an effective regulatory system gave rise to a classic collective-action problem in which individual whalers lacked any incentive to limit their own harvesting in the name of conservation even as the actions of all produced the dynamic we now call the "tragedy of the commons" (Baden and Noonan 1998).

Most early efforts to devise an effective management regime in this area emphasized conservation in the form of what some observers have labeled the "gospel of efficiency" (Hays 1975). These efforts focused on whales and sought to achieve "proper conservation of whale stocks and thus make possible the orderly development of the whaling industry" (ICRW Preamble). What we see here is an application of a familiar and influential paradigm dealing with efforts to conserve living or renewable natural resources. Commercial harvesting of animals, on this account, is legitimate and ethically or morally acceptable. But there is a need to regulate the practices of harvesters to avoid severe depletions, economic ruin, or even the extinction of some individual species of whales. The goal of governance in such settings is to manage harvesting of discrete stocks in such a way as to achieve and maintain maximum sustainable yields (MSY). This way of thinking does not pay attention to the roles the animals in question play in larger ecosystems or to the unintended effects of other activities (e.g., commercial shipping) on the condition of individual stocks. At the time of the adoption of the International Convention for the Regulation of Whaling in 1946, MSY constituted the prevailing discourse regarding the use of renewable resources; the provisions of the convention reflect this mode of thought.

The conservation discourse remains an influential frame of reference today, clearly visible in many debates about measures pertaining to whales and whaling. But two competing discourses have emerged that challenge the gospel of efficiency—and sometimes each other—as sources of guidance for those responsible for making decisions regarding human actions affecting whales. For shorthand purposes, we can think of these new approaches as the discourse of ecosystem-based management (EBM) and the discourse of preservationism.

Arising conceptually and analytically from the emerging field of conservation biology, ecosystem-based management provides a paradigm or way of thinking that researchers and practitioners have applied to a wide range of situations involving both consumptive uses of living resources and the unintended side effects of other activities (e.g., commercial farming) on living resources (Norse and Crowder 2005). The basic idea is that living organisms like whales are embedded in larger biophysical systems that must be taken into account in any effort to manage the harvesting of these resources. Changes in these larger systems (e.g., increasing land-based pollution, rising water temperatures, shifting ice conditions, and rising sea levels) can have far-reaching consequences for whales, over and above human harvesting. The harvesting of whales can also alter the dynamics of large socioecological systems. Both the complexity of these systems and their tendency toward nonlinear and often abrupt patterns of change make it hard to determine what level of harvesting can occur without triggering cascades of change that cause these large ecosystems to flip into different—and often undesirable—states (Walker and Salt 2006). EBM does not lead to the conclusion that we should terminate all harvesting of living resources; it does not license any particular position on issues like the debate over whether individual whales should be accorded a right to life. But it does call for caution in making decisions under uncertainty as well as for a consideration of the broader impacts on underlying ecosystems arising from activities like the harvesting of whales.

Preservationism directs attention to the treatment of individual animals. Those who subscribe to this paradigm care about the welfare of individual whales and focus on the extent to which we should accord them a right to life that trumps any right to harvest living organisms for human purposes. Preservationism sometimes appeals to ideas associated with ecosystem-based management. But it is not fundamentally an argument about conservation or the condition of complex ecosystems. Rather, it is a paradigm rooted in ethical principles as applied to interactions between humans and other living organisms (Singer 1975). Even within the realm of environmental ethics, preservationism adopts a distinct stance. It is not a systemic perspective of the sort exemplified by Aldo Leopold's familiar concept of the "land ethic" (Leopold 1970). It is a perspective that emphasizes the welfare of individual animals, especially those like whales and elephants that are large, charismatic, intelligent, and apparently threatened as a consequence of human actions (Freeman and Kreuter 1994). Unlike both conservationism and the idea of EBM, preservationism is associated with the view that the only defensible position for the International Whaling Commission is to oppose all harvesting or consumptive uses of whales.

The existing regime for whales and whaling is dominated by discord and sometimes open conflict among those whose thinking about the role of the IWC is rooted in one or the other of these perspectives. Many of those who supported the decision taken in 1982 to impose a moratorium on commercial whaling, for instance, made their case predominantly in terms of the tenets of conservation and EBM. Their goal was rebuilding whale stocks and devising the Revised Management Procedure (RMP), which would ensure the sustainability of harvesting following the lifting of the moratorium (Friedheim 2001a). But preservationists have claimed the moratorium as the moral high ground and refused to accept any resumption of harvesting, even under the exceptionally rigorous management procedures articulated in the RMP and operationalized in the Revised Management Scheme (RMS). Because decisions of the IWC regarding such matters require a three-fourths majority, it is difficult to gain agreement on any departure from the status quo. The moratorium, adopted in 1982 as a temporary measure, remains in place as the default option.

Today, the regime for whales and whaling is in a state of gridlock arising from confrontations among these dueling discourses. The regime exemplifies the pattern I call diversion in the sense that it is no longer dedicated to the goal articulated in the constitutive document—the 1946 ICRW—of introducing conservation measures to promote "the orderly development of the whaling industry" (ICRW Preamble). But a carefully crafted and broadly appealing alternative has not come into focus to carry the day. The preservationists are determined to maintain the moratorium at all costs. Yet some IWC members are harvesting whales justified either under the provision in Article VIII of the ICRW allowing for "scientific whaling" (e.g., Japan) or through reliance on the regime's objection procedure (e.g., Norway and Iceland). And the regime has not been able to broaden its scope to encompass many species of small cetaceans not included under the rubric of "great whales" in order to address threats to whales that do not involve harvesting (e.g., pollution, increased ship traffic, and underwater noise), or, more generally, to address the likely impact on whales of the overarching crisis in ocean governance (Crowder et al. 2006; Young et al. 2007). On a day-to-day basis, the result is a pattern of diversion followed by a condition of paralysis in which the major players have been unable to build a winning coalition needed to move forward in any discernible direction. These circumstances make it easy to fall into the habit of treating this condition of gridlock as a fact of life destined to remain with us for the foreseeable future. Yet this is exactly the sort of situation in which seemingly small changes can push a system over a threshold or past a tipping point and, as a result, trigger far-reaching and sometimes abrupt changes. It is hard to forecast if and when such a state change will occur with regard to the regime for whales and whaling. But it would be a mistake to ignore this possibility.

Facts: A Brief History of the Regime for Whales and Whaling

Well into the twentieth century, the management of whaling—like the management of most fisheries—rested on what is commonly known as the law of capture. Under this open-to-entry regime, whales were regarded as common property right up to the moment of capture, at which point they became the private property of those who captured them. Because whalers—as well as others who thought about the issue at all—assumed implicitly if not explicitly that human harvesting would not lead to serious depletions of marine living resources including whales, public authorities made little or no effort to impose entry barriers or to regulate harvesting methods in the interests of ensuring sustainable harvests. The fact that much of the harvesting took place in international waters beyond the reach of the regulatory jurisdiction of coastal states merely reinforced the propensity of governments to adopt a hands-off policy regarding the activities of whalers.

So long as the supply of whales was sufficient to withstand the pressures placed on them by whalers, this management strategy produced acceptable results, at least in terms of the tenets of conservationism as the principal source of environmental concerns. Despite the optimism of some scientists about the bountiful resources of the oceans, the fact that whale stocks were subject to more or less severe depletion as a consequence of human harvesting became known—at least among the whalers themselves—by the end of the eighteenth century (Dolin 2007). The story of whaling in the nineteenth century is, in large measure, an account of longer and longer voyages to more distant seas in pursuit of dwindling stocks of whales. By the second half of the nineteenth century, whalers were engaged in risky harvesting activities in the seas bordering the Arctic Ocean. The early years of the twentieth century witnessed an upsurge of whaling activities in the Southern Ocean surrounding Antarctica. The need for a more effective system of management was apparent even to advocates of laissez-faire practices by the early 1900s. Following a hiatus attributable to World War I, whaling resumed with a vengeance in the 1920s. The use of sophisticated harpoon guns and factory ships made it possible to kill large numbers of whales even in remote locations like the Southern Ocean. By 1925, the League of Nations had taken notice of the depletion of whale stocks and recognized the need for an international regime designed to manage the harvesting of whales on the high seas (Small 1971). This wave of concern led to the establishment of the International Bureau for Whaling Statistics in 1930, followed closely by the adoption of the Convention for the Regulation of Whaling in 1931. Although this convention was a landmark in the sense that it represented a conscious effort to regulate whaling on the high seas, the resultant regime lacked teeth and failed to make a dent in dealing with the problem of the depletion of whale stocks. The Great Depression of the 1930s followed closely by World War II led to the demise of this initial effort to create a management regime to regulate the activities of whalers.

As table 5.1 indicates, a new regime came into existence with the adoption in 1946 of the International Convention for the Regulation of Whaling (ICRW), which entered into force in 1948 and authorized the establishment in 1949 of the International Whaling Commission as the operational body responsible for making regular decisions about allowable harvest levels and authorized to adopt a variety of regulatory measures to sustain whale populations. Signed initially by fourteen countries, the ICRW articulates a management philosophy that joins a concern for conservation of whale stocks with the goal of managing continued harvesting of whales for human consumption on a sustainable basis. Thus, the ICRW preamble articulates "the interest of the nations of the world in safeguarding for future generations the great natural resources represented by the whale stocks" and calls for the achievement of "the optimum level of whale stocks as rapidly as possible without causing widespread economic and nutritional distress." The convention is explicit in calling for the "orderly development of the whaling industry" as a principal objective (ICRW Preamble). Like many regimes created to manage human uses of living resources during the early postwar era, the regime for whales and whaling focuses on the relevant stocks and features an effort to shore up the sustainability of harvesting to support consumptive uses on the part of humans. It does not embrace the idea of ecosystem-based management.

At the heart of the ICRW is a document known as the Schedule that serves as the principal vehicle for managing whale stocks on an ongoing

Table 5.1
Regime for Whales and Whaling Timeline

1820s–1850s	Golden Age of Whaling
1859	First oil well—Titusville, PA
1870–1880s	Whaling cannon/harpoon gun developed
1930	International Bureau for Whaling Statistics established
1931	International Agreement for the Regulation of Whaling adopted
1946	International Convention for the Regulation of Whaling (ICRW) adopted
1949	International Whaling Commission (IWC) established
1956	Protocol extending provisions to aircraft adopted
1961	Highest known number of whales killed (66,000)
1972	Use of blue whale unit (BWU) abolished
1972	UN Conference on the Human Environment
1975	IWC adopts New Management Procedure
1982	Moratorium on whaling to start with 1985–1986 season
1987	Japan begins scientific whaling
1989	Lowest known number of whales harvested (326)
1992	North Atlantic Marine Mammal Commission (NAMMCO) established
1993	Norway resumes whaling using objection procedure
1994	IWC adopts Revised Management Procedure (RMP)
1994	Southern Ocean Whale Sanctuary established
1990s	IWC works on Revised Management Scheme (RMS)
2002	Iceland rejoins with objection to the moratorium
2003	Conservation Committee established
2001–2008	ICRW membership doubles
2010	ICRW has 88 members

basis. Through amendments to the Schedule, the International Whaling Commission can act to establish annual quotas for the harvesting of whales and to employ a suite of regulatory devices made familiar from efforts to manage fisheries (e.g., open and closed areas, limited seasons, and gear restrictions). Through the 1950s and 1960s, levels of uncertainty regarding the condition of individual whale stocks remained high. Whaling nations reluctant to impose serious restrictions on their whalers dominated the IWC. The result was a string of decisions to set high quotas and to minimize restrictions on the activities of whalers. Stocks of great whales (especially blue, fin, right, and bowhead) continued to decline. Any objective assessment of the performance of this regime during these years would have to conclude that it had evolved into a whalers club unable to make the tough decisions about restrictions needed to rebuild stocks.

All this changed in the 1970s and 1980s (Stoett 1997; Andresen 1998). Due to an influx of nonwhaling nations and the cessation of commercial whaling by many of the original members (e.g., Australia, the Netherlands, the United Kingdom, and the United States), the regime for whales and whaling underwent a sea change. What had been a whaler's club now shifted step-by-step into a regime dominated by states having little or no interest in protecting whaling as an important industry. During the same period, a variety of nongovernmental organizations (NGOs) interested in whales came into existence, rose to positions of influence, and found the plight of great whales a powerful symbol of what was wrong with human-environment relations more generally during the closing decades of the twentieth century. These NGOs divided into two main groups with distinct agendas regarding the protection of whales. Some, like the World Wildlife Fund, were inspired by the ecosystem perspective articulated powerfully by publicists like Rachel Carson, whose highly influential book *Silent Spring* appeared in 1962 (Carson 1962); they brought to bear a heightened concern for the protection of ecosystems and for the roles that whales play as top predators in a variety of marine systems. Others, like the International Fund for Animal Welfare, adopted preservationist outlooks. Over time, it has become apparent that the views of supporters of these ecosystem-based and preservationist discourses differ on a number of important matters. But as the 1970s turned into the 1980s, the two groups found common ground in taking steps to halt the depletion of whale stocks and to turn the regime for whales and whaling into a mechanism for protecting and rebuilding depleted stocks.

The high-water mark of this coordination came in 1982 with the IWC's decision to impose a moratorium on the killing of great whales starting with the 1985–1986 season. Passed by the required three-fourths majority—made possible by shifts in the regime's membership together with conservationists' realization of the need for reform, facilitating an alliance between those favoring EBM and preservationists—the moratorium was presented at the time as a temporary measure needed to rebuild stocks and to devise revised management procedures capable of ensuring the sustainability of harvests once the moratorium was lifted. In the ensuing years, the impact of divergent or dueling discourses became increasingly apparent (Dryzek 1997). The preservationists saw the moratorium as an initial step toward a permanent ban on killing all great whales, to be followed by additional measures such as creating whale sanctuaries encompassing large segments of the world ocean. Supporters of EBM called for enhanced scientific efforts to provide the knowledge base needed to develop what became known as the Revised Management Procedure. Those who had espoused a conservationist perspective all along pointed out that some stocks of whales (e.g., minke whales in the North Atlantic) were large enough to sustain a commercial harvest, so long as it took place under the provisions of an effective regulatory regime.

The ensuing years have featured a pattern in which gridlock in the operation of the regime for whales and whaling has followed in the wake of the imposition of the moratorium. Membership in the regime has grown (there are now eighty-eight members) in a manner that is, on balance, somewhat favorable to the interests of whalers. But the array of views among the members is such that there is no prospect of mustering a three-fourths majority in support of any significant change in the status quo. Due to the nature of the regime's decision rules, this means that the status quo has become the default option. The moratorium, adopted in 1982 as a measure intended to last no more than ten years, is still in place. But other changes have occurred that make the status quo increasingly unappealing to a number of players in this regime.

The scientific community has concluded that limited commercial harvests of some whale stocks would not jeopardize either the whale stocks themselves or the ecosystems of which they are a part. On the strength of these findings, the IWC and its Scientific Committee have devised a system of highly sophisticated management procedures articulated in the RMP and RMS. Although the IWC accepted the RMP in 1994, those opposed to any killing of whales have blocked the implementation of these arrangements. Meanwhile, Norway and Iceland have resumed commercial whaling as they are entitled to do under the objection procedure set forth in Article IV(3) of the ICRW; Japan has resumed what amounts to commercial whaling under the "scientific research" provision of Article VIII. Norway, Iceland, the Faroe Islands, and Greenland have established the North Atlantic Marine Mammal Commission to address issues involving the harvesting of living resources on the basis of ecosystem-based management (Hoel 1993). The regime for whales and whaling has degenerated into an antiquated arrangement that has become a forum for dogged and inconclusive battles among those holding incompatible views not only on specific management procedures but also on the fundamental goals of efforts to manage human actions affecting whales and the ecosystems to which they belong.

This regime exemplifies the pattern of change I have called diversion. What was established as a mechanism for managing the harvest of whales on a sustainable basis has evolved into an institutional arrangement that puts significant impediments in the way of those, like the Japanese and the Norwegians, wanting to engage in whaling on the basis of procedures like those articulated in the RMP and RMS. It is hard to see any way forward regarding these issues, at least under prevailing conditions. The resultant gridlock has given rise to centrifugal forces that could precipitate a crisis leading to the collapse of the regime for whales and whaling as we have known it over the past twenty-five years. This makes it impossible for the IWC to deal effectively with the actions of Japan, Norway, and Iceland, much less to address the impacts of exogenous forces, such as increased ship traffic, noise pollution, and climate change, that are already affecting a number of whale stocks. The regime remains a relic of the era of management based on the idea of maximum sustainable yields; it has little capacity to adapt effectively to changing circumstances (Larkin 1977). Seemingly small perturbations can trigger the eruption of crises in situations of this sort. Such an event would be anxiety producing, at least to those who cling to the existing regime as a familiar arrangement, despite its increasingly obvious shortcomings. But a crisis might also provide an opportunity to introduce far-reaching and constructive changes needed to make this regime useful in the twenty-first century.

Analysis: Sources of Diversion

How can we explain the pattern of diversion followed by gridlock that is such a marked feature of the regime for whales and whaling, at least from the 1980s onward? Some of the relevant factors are easy to identify. Exogenous forces like the rise of antiwhaling sentiments in many quarters and the growth of influential NGOs capable of giving voice to these sentiments, for example, are important elements of this story. Yet it would have been difficult for these forces to make headway in shifting the center of gravity in the IWC in the absence of key features of the regime itself, or what I call endogenous factors. This section addresses the alignment of endogenous and exogenous factors in the case of whaling, starting as usual with the endogenous factors, moving on to exogenous factors, and concluding with a look at the alignment between the most important endogenous and exogenous factors.

Endogenous Factors

In some ways, the regime established under the terms of the ICRW was designed to fail, even when treated as a conservation arrangement dedicated to achieving sustainable harvests of various stocks of whales. A number of features of the regime itself contributed to this situation, including the role of the harvesters in setting total allowable catches (TACs) applicable to themselves, the lack of a strong and independent scientific committee capable of providing well-grounded advice on the status of whale stocks, the failure to agree on a method for allocating TACs among whalers or even among whaling nations, the denomination of TACs in terms of blue whale units (BWUs) rather than on a species-by-species basis, and the lack of provisions allowing for the development of effective compliance mechanisms.

A few observations regarding each of these concerns will serve to clarify the scope of the problem. Regimes in which the harvesters themselves set TACs are notorious for establishing quotas that are too generous from the perspective of conservation; the IWC proved to be no exception in these terms. Acting formally through a procedure involving annual amendments to the Schedule, the commission repeatedly established allowable harvest levels during the 1960s and 1970s that were not sustainable on an ongoing basis. The fact that these quotas were stated in terms of blue whale units and that no attempt was made to allocate quotas among the whalers or whaling nations simply added to the resultant problems. The use of the BWU—a formula under which 1 blue whale equaled 2 fin whales or 2.5 humpbacks or 6 sei whales—as the currency for computing the overall quota until 1972 made it impossible to devise management procedures tailored to the circumstances of individual species, much less individual stocks of whales (Friedheim 2001a). The absence of procedures for allocating quotas among harvesters gave rise to behavior that is all too familiar in connection with the consumptive use of living resources. Taking the form of what became known as "Whaling Olympics," the resulting competition among harvesters played out on an annual basis at the expense of the stated goals of the 1946 convention.

The problems of uncertainty and underdeveloped compliance mechanisms exacerbated this situation. The regime established under the terms of the ICRW does provide for the Scientific Committee (Aron 2001). But knowledge regarding the existence and condition of well-defined whale stocks was severely limited during the early years of this regime. One reasonable response to this situation would have been to adopt a highly conservative approach to setting quotas based on what we now know as the precautionary principle. But not only was this principle unknown at the time; the Scientific Committee also suffered from politicization, a condition that was hardly conducive to playing an effective role in regulating whaling in the interests of achieving sustainable yields (Andresen 1998). The absence of an effective compliance system intensified the resultant problems. The fact that most whaling took place in remote areas of the high seas made it virtually impossible to monitor the actions of whalers, much less to devise a system of sanctions capable of deterring prospective violators of the provisions of the regime. We now know that this was much more than a theoretical problem with regard to this regime. Persistent and severe underreporting of data on harvesting during the 1950s and 1960s on the part of the Soviet Union surfaced some years ago (Yablokov 1997; Morell 2009). Suspected by some scientists endeavoring to understand the dynamics of whale stocks, the actions of the Soviet Union in violating quotas established by the IWC came to light only in the 1990s; this revelation did not lead to any concerted effort to impose sanctions on the Russian Federation (as the successor to the Soviet Union), much less the officials responsible for these violations of the regime's provisions.

By the late 1970s, the circumstances prevailing in the early years of the regime for whales and whaling had changed dramatically. Many whale stocks were clearly in trouble; whalers commonly found it difficult to harvest enough whales to fill the quotas that the IWC set. A number of countries (e.g., Australia, the Netherlands, and the United States) that were original members of the ICRW had ceased to operate as whaling nations (with exceptions for aboriginal subsistence whaling). Several countries with no history of participation in the production and consumption of whale products had taken advantage of the ICRW provision allowing countries to join the regime without meeting any requirements for membership. There is some evidence that individual countries opposing the continuation of whaling were recruited as new members of the regime by those advocating a cessation of whaling (DeSombre 2001). The scientific community was increasingly skeptical of approaches to management based on the concept of MSY; perspectives that we now think of as ecosystem-based management were on the rise in this community. The antiwhaling forces based on preservationist perspectives were becoming stronger with each passing year. The preservationists had turned out in force at the 1972 UN Conference on the Human Environment. With the passage of time, the antiwhaling campaign mounted by the preservationists gathered public support and became a force to be reckoned with in this sphere.

Under the circumstances, the stage was set for the creation of a broad coalition concerned with the performance of the regime for whales and whaling and ready to take strong measures to bring about change. The ICRW sets the bar high for those campaigning for significant changes in the operation of the regime. Changes in the Schedule require a three-fourths majority, a provision that makes the status quo the default option under most circumstances (ICRW Article III). But with the onset of the 1980s, a powerful coalition of countries calling for change emerged. Professional managers concerned with sustainable yields were alarmed by evidence of the decline of many whale stocks. Scientists increasingly committed to the idea of ecosystem-based management came to see the provisions of the ICRW as part of the problem rather than part of the solution. The rising tide of preservationism produced an increasingly vocal and influential campaign calling for the cessation of intentional killing of whales in any form. This set the stage for the adoption of the moratorium on commercial whaling during the 1982 session of the IWC. In formal terms, the moratorium has set allowable harvests under the terms of the Schedule at zero. The motives of those who joined the coalition supporting the moratorium in 1982 were diverse, a condition that has affected the performance of this regime ever since. But this coalition was able to muster the three-fourths majority needed to adopt the moratorium in 1982 and, as a result, to make use of the voting rules embedded in the regime for whales and whaling to bring about a change in the regime that has had the effect of diverting this governance system from its original objective of making possible "the orderly development of the whaling industry."

Initially, advocates justified the moratorium as a temporary measure designed both to permit whale stocks to recover and to allow the IWC to develop more effective procedures for setting TACs and ensuring compliance on the part of whalers. The understanding among many who supported the moratorium was that it would last about a decade and be followed by the adoption and implementation of what has become known as the Revised Management Procedure. But this did not happen. More than twenty-five years later, the moratorium remains in place, and there is no immediate prospect of building a coalition in support of lifting the moratorium that can command the three-fourths majority required to adopt new procedures.

What accounts for this paralysis, and what are the prospects for the future? To a sizable degree, answers to these questions focus on specific features of the regime having to do with membership as well as the rules governing decision making in the IWC. Having accomplished its initial goal, the broad coalition that formed to adopt the moratorium in 1982 proceeded to split apart. On one side stood the advocates of ecosystem-based management, joined for the most part by the conservationists who now recognized that a simple return to a management procedure based on the idea of MSY was not a feasible option. Confronting them across a deep normative divide stood the preservationists who had no intention of agreeing to new procedures that would sanction intentional killing of whales from any stock or for any purpose.

Supporters of ecosystem-based management busied themselves with expanding scientific knowledge regarding whale biology and behavior and devoted their energy to the development of new and more effective methods of managing whale harvests that have taken the form of what we now know as the Revised Management Procedure and Revised Management Scheme. In some ways, the results of their efforts have been impressive (Aron 2001). Some whale stocks (e.g., North Atlantic minke whales) could sustain an annual harvest without depleting the stocks in question. Other stocks are clearly rebounding in response to the moratorium; some are reaching a stage where they could sustain modest and properly managed harvests. Equally important, the RMP sets forth a management regime that is widely regarded as one of the most sophisticated and cautious management systems pertaining to living resources in existence. It would make use of the precautionary principle in setting quotas and calls for procedures that would go a long way toward solving the problems of monitoring, reporting, and verification that plagued earlier procedures dealing with the harvesting of whales. The IWC actually voted to approve the RMP at its 1994 meeting. But without an agreement to set quotas through amendments to the Schedule, the RMP (along with the practical procedures set forth in the RMS) remains a paper arrangement.

Meanwhile, the preservationist forces took steps to protect and expand the gains they made in the 1980s largely through the adoption of the moratorium. These forces have enjoyed marked success in mobilizing public opinion around the world in opposition to the killing of whales. Within the IWC, they have fought for restrictive measures like the creation of large sanctuaries in which all whaling is banned. This has resulted in the (contested) creation in 1994 of the Southern Ocean Sanctuary; proposals to create additional sanctuaries have been on the table in recent years. At the same time, the preservationists have taken advantage of their numbers as regime members with voting rights to ensure that the moratorium remains in place and that the RMP and RMS for governing whaling do not become operational. The preservationist forces cannot muster the three-fourths majority needed to take additional steps toward the termination of whaling on a permanent basis. But they can block efforts on the part of any other coalition to move the RMP and RMS from paper to practice.

Several additional observations about the endogenous factors deserve emphasis in this discussion. The ICRW does not specify any qualifications for membership, and new members have continued to join the regime for whales and whaling. A sizable fraction of the new entrants are sympathetic to the views of whalers, especially those engaged in aboriginal subsistence whaling or small-type coastal whaling. The preservationists allege that Japan has actively recruited some of the new members with various types of bribes, and the prowhaling forces have done better in IWC voting in recent years. These forces even succeeded in mustering a simple majority in a vote taken during the 2006 IWC meeting dealing with a matter of some importance. Still, it is hard to imagine the prowhaling forces commanding a three-fourths majority in the IWC within the foreseeable future. The rules of the regime give each side veto power over the proposals launched by the other.

Those in favor of harvesting whales have managed to take advantage of several provisions of the regime to carry out whaling on a limited but significant basis. Norway filed an objection to the 1982 moratorium under the provisions of Article V(3) and is therefore not subject to either temporary or long-term bans on the harvesting of whales. Japan, which originally filed an objection but later withdrew it, now harvests a sizable number of whales using the provisions in Article VIII of the ICRW allowing the parties to kill whales for purposes of scientific research.

Iceland succeeded in getting itself readmitted to the regime in 2002 with a reservation to the moratorium intact. Altogether, the resultant harvests now encompass the killing of two thousand or more animals per year. The upshot is a situation in which the moratorium itself is a sacred cow, but whalers have found ways to carry out limited harvests of whales by taking advantage of mechanisms that are integral to the regime and therefore properly treated as endogenous factors.

Well-informed observers acknowledge that the regime for whales and whaling is in disarray (Friedheim 2001c). To a considerable extent this is due to features of the regime itself, including those pertaining to the entry of new members, the voting rule regarding important matters, and the reservation procedure. Even so, the regime has proved sticky, or highly resistant to proposals for serious reform. Many members have invested considerable time and diplomatic capital in crafting proposals to overcome the current deadlock. The so-called Irish proposal, first tabled in the late 1990s by Ireland and calling for agreement on a package of reforms including measures of interest to various parties (e.g., designating a global whale sanctuary and allowing for a resumption of small-type coastal whaling), for instance, has garnered support among a sizable fraction of the regime's members (Burke 2001). But efforts to assemble a three-fourths majority have failed in every case. The decision rule calling for a three-fourths majority guarantees inaction whenever and so long as the members of the regime remain deeply divided regarding both the ends to be pursued and the means to be used in applying the provisions of the regime to real-world situations.

Exogenous Factors

External factors affecting the regime for whales and whaling have evolved over the period analyzed in the preceding subsection. At the leading edge of these developments comes change in the attitudes of the public toward the practice of killing whales for human consumption. Through at least the early decades of the twentieth century, public attitudes toward the harvesting of whales were positive. Most looked upon whaling as part and parcel of fishing more generally, a way of approaching the issue that tended to suppress doubts about the acceptability or legitimacy of killing whales on normative grounds. The shift in attitudes during the past fifty years has been remarkable. Propagated by a variety of NGOs (e.g., Defenders of Wildlife, Greenpeace, the Humane Society, and the International Fund for Animal Welfare), the view that whales are sentient beings and that killing them for human purposes is unethical has spread far and wide among members of the general public. The rise of the animal rights movement from the 1970s onward to a position of considerable influence has strengthened the hand of those calling for a permanent ban on killing whales in the setting of the IWC (Singer 1975). This movement may well continue to grow, following the pattern exemplified by the antislavery movement in the nineteenth century and the civil rights movement in the twentieth century. But continued growth of the antiwhaling movement is not inevitable. Some movements rise to a certain level and then recede without fulfilling or solidifying progress toward their stated goals. The prohibition movement in the United States during the first half of the twentieth century is a case in point. We should not treat it as an article of faith that the practices of prowhalers are destined to fall by the wayside as we move deeper into the twenty-first century.

Two other external forces lend weight to the view of those who believe that the antiwhaling forces are destined to grow and ultimately to win the battle to end the killing of whales. Markets for whale products have largely dried up over the course of the past century and especially over recent decades. Cheaper and better substitutes for whale oil are now widely available, and simple substitutes for most other whale products (corsets, candles, and so forth) are in general use. Whereas whaling was once one of the top industries in a number of countries—within the top five in the United States during the first half of the nineteenth century (Dolin 2007)—the industry is now marginalized. It is impossible to devise credible scenarios featuring a reversal of this trend in the foreseeable future. Several countries (e.g., the United States) have dropped out of the ranks of the prowhaling coalition for reasons having little or nothing to do with the activities of the regime created under the terms of the ICRW. There is no effective interest group or lobby capable of protecting today's whalers in the policy-making process in most countries. The rising tide of preservationism in many circles has generated a set of circumstances in which the remaining political clout of defenders of whaling is largely drowned out in the chorus of voices calling for policies that accord individual whales a right to life, at least with regard to intentional killing on the part of various human groups still engaged in some form of whaling.

In this setting—featuring a rising tide of prolife sentiments regarding the fate of whales and a steep decline in the whaling industry—the proliferation of NGOs and the emergence of what many observers describe as global civil society loom large as external forces that have a bearing on the regime for whales and whaling (Peterson 1993; Wapner 1997). We now live in a world featuring a rapidly growing array of nonstate actors capable of exercising significant influence in policy-making processes. Some of these actors (e.g., the World Council of Whalers and the High North Alliance) are strong supporters of—mostly aboriginal or small-type coastal—whaling; they are a force to be reckoned with in a number of forums, such as the North Atlantic Marine Mammal Commission and the Arctic Council as well as the IWC itself. But far more numerous and well funded are the NGOs dedicated to the proposition that intentional killing of whales must stop. These groups have articulated a variety of views regarding whaling. Some mainstream environmental groups, including the Worldwide Fund for Nature and even Greenpeace, have adopted cautious views on this topic and lent support from time to time to those engaged in aboriginal subsistence whaling. But a large swath of animal rights groups, including the Animal Welfare Institute, Defenders of Wildlife, the Humane Society, and the International Fund for Animal Welfare, have pursued the goal of terminating the intentional killing of whales with religious fervor. Often well funded and populated with people who are sophisticated users of modern information technologies, these groups commonly argue that they are acting in the public interest and are not interest groups in the traditional sense of the term. The antiwhaling movement has drawn strength from its association with the rise of broader sentiments favoring the acknowledgment of animal rights. Charismatic megafauna, including whales as well as elephants and tigers, have played important roles as the poster species for the animal rights movement (Freeman and Kreuter 1994). The plight of whales has bolstered this movement, just as the movement itself has become a more and more effective force calling for the adoption of a comprehensive and permanent ban on the intentional killing of whales.

Several factors of a more political nature also deserve attention in this discussion of external determinants of the emergent pattern of diversion followed by gridlock in the regime for whales and whaling. One centers on the capacity and the desire of key actors to deploy financial incentives to encourage like-minded states to become members of the regime. Because Article X of the ICRW allows any state to become a member without fulfilling any specific requirements for membership, an opportunity for manipulative intervention arises. The cost of membership in this regime—essentially the cost of representation at annual meetings of the IWC—is modest. States, including landlocked states like Switzerland, can and do choose to become members even when they do not have large stakes in the relevant issues. This means that current members or even well-funded NGOs seeking to form winning coalitions within the IWC or to block the formation of such coalitions on the part of others can make use of economic incentive to recruit new members likely to share their views (DeSombre 2001). In extreme cases, they may even act informally to defray the expenses incurred by new members. There is evidence to suggest that proponents of the moratorium engaged in such efforts in the 1970s and early 1980s. A number of observers have alleged that Japan has acted in an analogous manner in recent years in an effort to win support for the adoption and implementation of the RMP and RMS. But the circumstances and, therefore, the consequences differ in the two cases. The effort to build the coalition needed to adopt the moratorium eventuated in success in 1982. More recently, the regime's membership has shifted in the direction of the preferences of Japan and other prowhaling members. But the prospect of building a coalition of three fourths of the current eighty-eight members in favor of implementing the RMP and RMS is remote.

Another political factor centers on efforts by individual members—in contrast to the regime itself—to bring pressure to bear on member states to alter their preferences and their behavior with regard to issues arising in the IWC. The prime examples here involve actions on the part of the United States to pressure regime members to adhere to the provisions of the moratorium and to desist from the continuation of whaling in any form (DeSombre 2000). U.S. legislation (e.g., amendments to both the Fishermen's Protective Act and the Fishery Conservation and Management Act) allows for the imposition of sanctions on any state whose actions diminish the effectiveness of the regime. According to many advocates, this applies both to the Norwegian harvest justified on the basis of filing a reservation to the 1982 moratorium and to the Japanese harvest justified under the scientific research provision of Article VIII of the ICRW. The counterfactual in this connection is anything but clear. It seems probable that the three-fourths majority needed to adopt the moratorium in 1982 would have emerged without such political pressure on the part of the United States. Similarly, it is hard to make the case that American pressure has played an essential role in preventing the development of the three-fourths majority needed to adopt and implement the RMP and RMS in recent years. But this case does illustrate the general proposition that regimes are subject to various forms of external political pressure, whatever the content of their formal rules and decision-making procedures. Such pressure almost always forms only a single element in a complex mix of forces affecting the fate of individual regimes. But that does not make it unimportant.

A different story emerges when we turn to an examination of developments in other spheres or issue areas that are likely to impinge on the conservation of whale stocks. The ICRW is a product of an era dominated by concerns about the prospect of depletions caused by overharvesting specific species or stocks. The prime concern of the arrangement created under the terms of the convention is achieving sustainable yields from harvested stocks. Yet whales, like other living resources, are located within complex and changing systems whose dynamics may have far-reaching impacts on the status of whale stocks in the future. Increases in ship traffic can affect whales by increasing the risk of collisions and by raising the volume of underwater noise affecting the guidance systems of whales. Pollutants that drain into the sea from rivers and lead to the development of dead zones in the oceans can have detrimental impacts on habitats important to whales. Even whale-watching, which has become a sizable industry in some places, can be harmful to whales when the density of whale watchers becomes too great or whale-watching boats are not handled with care. It would make sense for a regime dealing with whales to play some role in identifying harmful effects of such actions and in taking steps to expand the scope of the regime to address these problems. But the existing regime is ill-suited to address such matters. This is a consequence, in part, of the emphasis of the ICRW on calculations of sustainable harvest levels. But more important, it is a result of the paralysis of the regime arising from the unresolved confrontation between those prepared to accept the harvesting of whales under certain conditions and those committed to the idea of granting individual whales a right to life. Externalities of the sort under consideration here may well emerge as greater threats to the welfare of whales than a highly controlled harvest of selected types of whales. But there is little chance that the existing regime can expand to cover these concerns.

Finally, there is the question of institutional interplay, or interactions between the regime for whales and whaling and other institutional arrangements in place today. Such interactions may take several forms. A number of whale stocks are listed in Appendix 1 of the Convention on International Trade in Endangered Species of Flora and Fauna (CITES). Hoping to pave the way toward lifting the moratorium, prowhaling forces have promoted down-listing some stocks under CITES. But these efforts have made little headway (Friedheim 2001b). There is also a link between the ICRW and the Convention on the Conservation of Antarctic Marine Living Resources (CCAMLR), not only because CCAMLR covers a sizable segment of prime whale habitat but also because this regime is an innovative arrangement with regard to the introduction of ecosystem-based management as an alternative to the traditional emphasis on sustainable yields. But conscious efforts to coordinate the work of the two regimes have been minimal.

The regime for whales and whaling is situated within the broader framework of the UN Convention on the Law of the Sea, opened for signature in 1982 more or less simultaneously with the adoption of the moratorium on whaling. Given that the regime for whales is constrained by the fact that it is a traditional conservation arrangement, there is much to be said for adopting a policy of linking this regime to newer governance systems in the interests of modernizing the regime established under the terms of the ICRW. But here, too, progress has been hampered severely by the atmosphere of gridlock that has plagued the regime for whales and whaling for many years.

Endogenous-Exogenous Alignment

How do the factors discussed in the previous subsections align with one another and what are the implications of these circumstances for the pattern of diversion and gridlock that marks the regime for whales and whaling? Conditions of alignment (or misalignment) have played a central role in the development of this regime, allowing first for the development of a winning coalition in support of the moratorium and later serving to block efforts to restructure the regime to allow it to preside over tightly controlled harvests under the terms of the RMP and RMS. Provisions of the regime have both allowed for diversion with regard to its basic goals and impeded efforts to swing back toward the original goal of the regime to "achieve proper conservation of whale stocks [thus making] possible the orderly development of the whaling industry."

The moratorium, presented initially as a temporary measure, garnered support from a broad coalition of conservationists, advocates of ecosystem-based management, and preservationists. All three groups had compelling reasons to find the regime's performance inadequate during the 1960s and 1970s and therefore to take advantage of the three-fourths decision rule to achieve a departure from business as usual in the form of the moratorium. The facts that several countries were in the process of abandoning whaling at the time and that many participants were no longer satisfied by the discourse of maximum sustainable yields played important roles in preparing the way for the adoption of the moratorium. So also did the influx of new members opposed to whaling.

Some of the same factors are relevant to explaining the inability to move beyond what was promoted initially as a temporary moratorium, thereby perpetuating the pattern of institutional change I call diversion. What has developed since 1982 is a persistent deadlock between those forces prepared to approve a resumption of whaling, so long as it occurs under the tightly controlled arrangements envisioned in the RMP and RMS, and the vocal opponents who are committed to according individual whales a right to life and who oppose any resumption of the harvesting of whales anywhere and under any conditions. The influx of new members in recent years—possibly supported by the Japanese in some cases—has tended to strengthen the hand of the prowhalers in the deliberations of the IWC. In 2006, the prowhalers even achieved a simple majority in favor of a measure calling for the resumption of whaling under controlled circumstances. But they did not come close to assembling the three-fourths majority needed to terminate the moratorium and replace it with the RMP and RMS or some similarly rigorous arrangement. There is little likelihood of breaking this stalemate any time soon. The result is a situation that not only fits the pattern I call diversion but also accounts for the inability of the regime to address a variety of issues (e.g., the consequences of increases in ship traffic, the growth of land-based pollutants affecting marine systems, or the impacts of climate change) that are likely to have more far-reaching consequences for the health of whale stocks in the future than a limited and strictly controlled harvest would have. This is not to say that the preservationists are wrong to espouse views that make it impossible to administer the regime in a manner that conforms to the intentions of those who crafted the terms of the ICRW. But the gridlock is not likely to end soon; the combination of external fluidity and internal rigidity makes it difficult to address serious concerns about threats to the well-being of a growing number of whale stocks.

Forecast: The Road Ahead

There is a lively debate regarding the extent to which the regime for whales and whaling has proven effective or successful. Several arguments are possible; none is correct in any objective sense. It is worth noting at the outset that many stocks of great whales are recovering from their depleted status during the dark days of the 1960s and 1970s. Whether these stocks are becoming robust, much less reaching their preharvesting levels, is a controversial matter. So also is the question of whether the creation and implementation of the regime played a significant role in fostering the recovery of particular whale stocks. There is a prima facie case for the proposition that the regime has made a difference in these terms. But since we cannot reach clear conclusions about the nature of the counterfactual in this realm, a note of caution is in order in addressing such questions.

Beyond this lie differences attributable to divergent views regarding the goal of the regime. Those, like the Japanese and the Norwegians, who take seriously the goal of making possible the "orderly development of the whaling industry" tend to conclude that the regime is a failure (Andresen 2002). They treat diversion as tantamount to the hijacking of the regime by those who have little or no interest in the survival of the whaling industry. Those who espouse ecosystem-based management or support the ideals of preservationism, on the other hand, have mixed feelings. The preservationists have succeeded in terminating conventional industrial whaling, at least for the foreseeable future. But they have not succeeded in reforming or even replacing the ICRW in formal terms, and they have not been able to terminate all killing of whales. Upward of two thousand animals a year are now killed in aboriginal subsistence whaling, scientific whaling, and whaling on the part of those not bound by the moratorium (see table 5.2). Equally important, those opposed to killing whales have not been successful in extending the reach of the IWC to encompass many species of smaller cetaceans, and they have not found ways to address newly emerging threats to the welfare of whales through the operation of the existing regime.

Advocates of ecosystem-based management can feel only frustration at the current state of affairs. Their views are often marginalized in the confrontation between consumptive users and preservationists. In the RMP and RMS, they have crafted what many regard as a highly sophisticated arrangement grounded in the tenets of ecosystem-based management. Also, they have made careful calculations regarding the extent to which individual whale stocks can sustain limited harvesting without causing harm to the stocks. But their views are frequently treated with suspicion by the preservationists who regard any management system that sanctions killing as wrong on ethical or moral grounds or they are sidelined in heated debates about the acceptability of killing whales under any circumstances. The idea of ecosystem-based management has become popular in scientific circles, and it is gaining credence among professional managers. But experience in the case of whales and whaling makes it clear that there are larger forces at work in the policy process

Table 5.2
Whaling by IWC Member Nations in 2004–2005 and Summer 2005

Area	Blue	Fin	Sei	Bryde	Sperm	Minke	Others	TOTAL
Southern Hemisphere:								
Japan	–	–	–	–	–	441[1]	–	441
North Atlantic:								
Denmark (Greenland)	–	13[2]	–	–	–	180[2]	–	193
Iceland	–	–	–	–	–	39[3]	–	39
Norway	–	–	–	–	–	639[4]	–	639
St. Vincent & the Grenadines	–	–	–	1[5]	–	–	1[5]	2
North Pacific:								
Japan	–	–	100[6]	50[6]	5[6]	222[6]	–	377
Korea	–	–	–	–	–	3[7]	–	3
Russian Fed.	–	–	–	–	–	–	126[8]	126
U.S.A.	–	–	–	–	–	–	68[9]	68
TOTAL	–	13	100	51	5	1524	195	1888

1. Special permit, including 1 struck and lost.

2. Aboriginal catch: 13 fin whales (including 1 struck but lost), 176 minke whales (including 3 struck but lost) in West Greenland, and 4 minke whales in East Greenland.

3. Special permit, including 5 struck and lost.

4. Commercial operation based on legitimate objection to the moratorium (including 6 whales struck but lost).

5. Aboriginal catch of 1 humpback whale, and illegal catch of 1 Bryde's whale.

6. Special permit, including 2 minke whales struck and lost.

7. Illegal catch of 3 minke whales.

8. Aboriginal catch of 2 bowhead whales, and 124 gray whales (including 9 struck and lost).

9. Aboriginal catch of bowhead whales, including 13 struck and lost.

References: Catches by IWC member nations in the 2004 and 2004–2005 seasons, prepared by the Secretariat (IWC/57/28). Catches by IWC member nations in the 2005 and 2005–2006 seasons, prepared by the Secretariat (Annex I of Chair's Report of the 58th Annual Meeting).

Source: Whaling Library Web site by Masaaki Ishuda, http://luna.pos.to/whale; table at http://luna.pos.to/whale/sta_2005.html.

that can and often will trump recommendations based on the application of the idea of ecosystem-based management.

What can we say about the future of the regime for whales and whaling? There are two scenarios to consider in responding to this question, one emphasizing stasis and a second pointing to the prospect of some sort of state change in the prevailing regime. The two scenarios are not altogether incompatible. Even in a setting characterized by stasis, some occurrence that seems minor in its own right can precipitate a cascade of developments leading to a state change.

Because environmental and resource regimes, like all other social institutions, are sticky, the default option in thinking about the future is always to expect the continuation of the status quo. Hysteresis is an important feature of regime dynamics, even in cases where regimes are quite dysfunctional or ill prepared to deal with a variety of challenges that have already come into focus or that are visible on the horizon. In the case of the regime for whales and whaling, a situation of this sort has prevailed for at least the past ten years (Friedheim 2001a). The IWC, the principal decision-making body of the regime, meets once a year. These meetings are characterized by more or less acrimonious encounters among the major factions that can be characterized loosely as the preservationists, the supporters of the RMP and RMS, and the defenders of aboriginal subsistence whaling. None of these factions is close to being able to muster the three-fourths majority needed to secure approval of its programmatic preferences. This does not preclude certain new initiatives, like the creation in 2003 of a conservation committee designed to draw attention to threats to whales that are unrelated to issues of harvesting. But it does ensure a condition of paralysis with regard to the resolution of the underlying issues facing the regime.

So far, the members of the regime have exhibited a remarkable willingness to live with this situation, despite its unsatisfactory character for all concerned. The Japanese have slowly increased the size of their scientific whaling program; the Norwegians have upped the ante regarding the harvesting of minke whales in the North Atlantic. For their part, the preservationists continue to press for the expansion of whale sanctuaries and oppose proposals to permit a resumption of small-type coastal whaling. Most whale stocks are in better shape than they were at the time of the adoption of the moratorium in 1982. All told, however, the harvest of whales now involves upward of two thousand animals per year, and the regime is unable to address emerging threats to whales that do not stem from intentional harvesting.

Why do the members of the regime put up with this frustrating situation? The simple answer is probably that no one can come up with a better alternative that is remotely realistic in political terms. A decision by the Japanese to resume open commercial whaling, for instance, would ignite a firestorm of criticism and could well trigger sanctions on the part of powerful players like the United States. For their part, the preservationists have few options, at least under current conditions. They do not have the votes to push through a permanent moratorium, and any effort to establish a new arrangement that they could dominate would likely lead to actions on the part of the Japanese, the Norwegians, and others that would only make matters worse from their point of view. The defenders of aboriginal subsistence whaling are reasonably satisfied with the status quo, though they know that the coalition backing their position is fragile. The members of the regime for whales and whaling remain engaged, in large measure, because they have no realistic alternatives and because they are loath to accept the political consequences likely to flow from a complete breakdown or collapse of the existing arrangements.

Situations of this sort do occasionally implode, an observation that brings me to the second scenario regarding the future of this regime. State changes occur infrequently. More often than not they are unforeseen, even by those who should be well informed about the forces at play. The collapse of the Soviet Union in the late 1980s and early 1990s, for instance, took even veteran Kremlin watchers by surprise. Additionally, seemingly small events that would not have precipitated far-reaching changes in earlier times can become triggers initiating cascades of changes in cases where a regime has lost resilience for one reason or another. As the case of the regime for northern fur seals (to be examined in chapter 6) suggests, a regime that has been around for a long time and that is widely admired by its supporters can collapse like a house of cards as a consequence of shifts in either endogenous or exogenous factors critical to its survival. Could some such transformation occur in the case of the regime for whales and whaling? I believe the answer to this question is in the affirmative; any of several triggers could destabilize the sticky but unsatisfactory state of affairs prevailing today.

Scientifically credible reports of sharp declines in important stocks of great whales could cause the existing regime to collapse. Despite its glaring shortcomings, this regime has played some role in the revival of a number of whale stocks from their low points in the midtwentieth century. Other factors have contributed to these positive developments, and any future declines would involve a multiplicity of causal mechanisms as well. But evidence of gains in key stocks of whales constitutes the current regime's ace in the hole. A reversal of these trends would almost certainly undermine the regime, regardless of the nature of the forces giving rise to such a situation.

A walkout by one or more members could also trigger the collapse of the existing regime. The obvious candidate for this role is Japan, a regime member that is increasingly frustrated by the inflexibility of the preservationists and that has resorted to questionable tactics to keep a modest whaling operation going (Friedheim 1996). There are good reasons for Japan to remain a member of this regime, not the least of which is the prospect of sanctions imposed by the United States in the wake of withdrawal. Still, such a walkout could occur. The Japanese have worked hard to build support for the implementation of the RMP and RMS. Although they have succeeded in constructing a sizable coalition in favor of this initiative, there is little likelihood of putting together the required three-fourths majority needed for the IWC to adopt this proposition. We cannot rule out a situation in which an increasingly frustrated Japan would opt for abandoning the current regime in favor of some alternative founded on the principles of ecosystem-based management. Such a move would almost certainly bring down the existing regime in short order.

Yet another possibility centers on growing threats to whales arising from human actions that the existing regime is unable to address in a meaningful fashion. The rise in incidents involving ships striking whales, increases in underwater noise, and the growing impacts of land-based pollutants that find their way into the sea could all play a role in this scenario. The regime for whales and whaling is a product of earlier thinking focused on achieving maximum sustainable yields; there is little chance that it can expand its coverage to include a range of externalities of the sort under consideration here. The emergence of issues of this kind as priority items on policy agendas would have the effect of marginalizing the existing regime. Should these concerns become long-term priorities in efforts to secure the survival of whales, the existing regime might simply fade away or become a dead letter based on a convention that remains in force legally but that is largely ignored by all relevant parties.

Should any of these scenarios—or others featuring conditions that might lead to a collapse of the existing regime—come to pass, what would happen next with regard to governance arrangements relating to whales? It is easier to say what would not happen than to forecast what would happen. There is no chance, for instance, that a successor regime would rest on a conservationist discourse with its emphasis on efforts to maximize yields from stocks of individual species. The heyday of thinking in terms of sustainable yields is over (Larkin 1977); there is no prospect of its revival under the circumstances prevailing today.

My expectation is that efforts to agree on a successor regime would feature a contest between the preservationists and those committed to the tenets of ecosystem-based management. All parties concerned would want to expand the coverage of the regime to encompass additional species of cetaceans, to think in terms of complex ecosystems in contrast to the role of individual species, and to devise stronger arrangements dealing with monitoring and enforcement. But those who think in terms of ecosystem-based management are not adverse to human harvesting of living resources so long as stocks are abundant and well-managed harvesting is a realistic option. They would likely push for final adoption and implementation of the RMP and RMS, a management arrangement that is widely regarded as state of the art from an EBM perspective. But this approach would not satisfy the preservationists. No doubt there is considerable variation among those who identify with this point of view. Still, all preservationists are committed more or less strongly to the proposition that individual whales as sentient beings should be accorded a right to life. Even the most well managed harvest imaginable would not be acceptable to those who espouse this point of view.

In most issue areas, I would expect ecosystem-based management, a way of thinking that recognizes the need to consider the dynamics of socio-ecological systems, to prevail in such a contest. But the status of whales as charismatic megafauna raises questions about the ability of ecosystem-based management to carry the day in this case. Public opinion is aroused easily with regard to specific incidents like the effort to liberate a few whales trapped in the ice north of Barrow, Alaska, or the Free Willy movement aimed at liberating a killer whale held in captivity. The tenets of ecosystem-based management do not lend themselves to presentation to the general public in the form of sound bites. Still, the preservationists might not welcome the collapse of the existing regime. Such a collapse might lead to some sort of hybrid arrangement incorporating elements from both discourses and producing outcomes that no one would find attractive.

One lesson arising from this account focuses on the need to think hard and creatively about the character of a successor regime well before an existing regime collapses. Windows of opportunity for major shifts in institutional arrangements do not stay open for long. Some alternative is likely to emerge in short order, whether or not it makes sense as a governance system for the situation at hand. Many well-informed observers have points of view regarding the deficiencies of the current regime for whales and whaling and about desirable reforms. But serious thinking about the features of a realistic replacement regime remains in short supply.

Conclusion

The dynamics of the regime for whales and whaling, one of the first experiments with environmental governance on a global scale, offer a clear example of the emergent pattern I call diversion. Created as an arrangement dedicated to the regulation of harvests to achieve the goal of maximum sustainable yields, the regime gave rise to a suspension of commercial whaling with the adoption of the moratorium in 1982. This is a case of diversion that has led to gridlock rather than to the development of some alternative framing of the fundamental goal of the regime. The regime remains the focus of a contest between preservationists who oppose the killing of whales under any circumstances and those who would accept some harvesting of whales so long as it is compatible with the tenets of ecosystem-based management. Diversion in this case was made possible by unusual circumstances permitting the creation of a winning coalition able to take advantage of the provisions of the ICRW allowing decisions on substantive matters to be made by a three-fourths majority. Ironically, the same provision has prevented the parties from lifting the moratorium as part of an agreement on the terms of a substantially reformed regime. The shifting composition of the regime's membership facilitated the creation of a winning coalition in support of the moratorium in 1982, but the prospects of fulfilling the three-fourths majority requirement for any package of reforms today are slim.

The result is a regime seemingly frozen in an awkward state of limbo. This situation may continue indefinitely. But there is also the prospect that a specific event (e.g., a decline in important whale stocks or a decision from Japan to abandon the prevailing regime) could push this system past a tipping point and bring the existing regime down with a crash. It is hard to forecast whether and when such a far-reaching change will occur. But there is a good case for taking this possibility seriously and investing time and energy in thinking about options for arrangements that could replace the existing regime and that would be better suited to the realities of the current era. There is a lack of such thinking in this case that is free of the ideological blinders that are common to policy analysis in a variety of settings.

6

Collapse: The Regime for Northern Fur Seals

Overview: The Big Picture

The regime for northern fur seals (*Callorhinus ursinus*), an early case of organized international cooperation designed to conserve wildlife, was widely admired—at least during some stages of its existence—as one of the most successful arrangements of its type. Yet this regime, formed initially under the terms of a four-party treaty signed in 1911, collapsed in the 1980s. The regime proved to be robust but not resilient in the face of major biophysical and socioeconomic changes. Once it reached a decisive tipping point, the regime disappeared without a trace.

The fur seal regime experienced two distinct cycles between 1911 and 1985. The first cycle, based on the 1911 Convention for the Preservation and Protection of Fur Seals, came to an end with the onset of World War II, a conflict that pitted one of the members of the regime—Japan—against the others—Russia (then the Soviet Union), the United States, and Canada. The signing in 1957 of the Interim Convention for the Conservation of Northern Fur Seals initiated a second cycle of this regime on terms closely resembling those of the first cycle. In 1983, the U.S. Congress passed a set of amendments to the Fur Seal Act of 1966, an initiative designed to discontinue the federal government's Pribilof Islands Program and, as a result, to terminate the harvesting of fur seals as a state-owned enterprise. Despite this fundamental change in American policy, the four parties to the 1957 convention proceeded in 1984 to negotiate the terms of a new protocol extending the life of the regime for several more years. But by then, the era of conservation arrangements based on the goal of maximizing sustainable harvests from individual species had run its course. The U.S. Senate chose not to ratify the 1984 protocol. Since the participation of the United States was (for reasons that will become apparent as this chapter progresses) essential to the operation of the system, the fur seal regime collapsed like a house of cards. By 1985, this venerable regime had vanished.

It is tempting to approach the problem that resulted in the creation of the fur seal regime as a straightforward matter of introducing rules to regulate the harvesting of what would otherwise be a common-pool resource, thereby avoiding the onset of what has become known as "the tragedy of the commons" (Hardin 1968; Baden and Noonan 1998). But the circumstances surrounding the harvesting of fur seals were both more complex and more dynamic than this simple way of framing the problem suggests. This is what makes the fur seal story especially interesting from the perspective of an analysis of institutional dynamics, even though the regime itself has been defunct since the 1980s.

Northern fur seals are migratory animals. Born on a few islands located mostly in the Bering Sea—the Pribilof Islands under the jurisdiction of the United States and the Commander Islands under Russian jurisdiction—fur seals migrate far to the south during the winter months along the coasts of North America and eastern Asia. Theoretically, it would be possible to harvest northern fur seals anywhere along these migratory routes. But such a practice would be highly inefficient and unprofitable. During the summer months, fur seals congregate on the Pribilof and Commander islands in the Bering Sea, giving birth to their pups and foraging for food in areas surrounding their rookeries. During this time, it is a simple matter to harvest seals on land, selecting bachelor males to minimize the impact of the harvest on the population as a whole. It is also possible—though inefficient and far less compatible with the requirements of conservation—to engage in pelagic sealing, taking animals while they are in the water but still close to their rookeries. Some interested parties have maintained that fur seals are available for anyone to harvest, so long as they are in the water in areas outside the jurisdiction of the coastal states. But seals harvested on land are not available for others to harvest so long as they are located on islands under American or Russian jurisdiction or in adjacent waters subject to the jurisdiction of these coastal states. The creation of Exclusive Economic Zones during the 1980s effectively provided the United States and Russia with the authority to control all harvesting of fur seals except during their migratory phase. It is easy to grasp the central tension embedded in this case between Russia and the United States as coastal states and Canada and Japan as distant-water states with a long-standing interest in harvesting this valuable living resource beyond the reach of coastal state jurisdiction.

Equally interesting is the organization of the harvest of fur seals, especially in the stronghold of these animals on the beaches and adjacent waters of the Pribilof Islands (Gay 1987). In the years following the purchase of Alaska in 1867, the United States leased the privilege of harvesting fur seals to private companies for periods of twenty years at a time. In conjunction with the effort to establish an international regime for fur seals early in the twentieth century, however, the federal government of the United States assumed responsibility for the management of the largest fur seal herds and proceeded to administer the harvest as a state-run enterprise (Young 1981a,b). The agency responsible for the Pribilof Islands Program—in later years the National Marine Fisheries Service—not only made decisions on an annual basis about allowable harvest levels but also organized and carried out the harvest, administered the sale of seal skins, and reaped the profits arising from this state-run industry. Although miniscule in relation to the U.S. federal budget, this income played a role of some importance as the principal indicator of a healthy industry. The harvesting of fur seals also became the economic mainstay of the Aleut communities of St. Paul and St. George, located on the two principal islands in the Pribilof group. Much has been written about various aspects of this practice, with some writers depicting the circumstances of the people involved as de facto slavery (Torrey 1978), while others viewed the situation as a form of benevolent paternalism. The passage of the Alaska Native Claims Settlement Act in 1971 brought major changes both in land ownership and economic organization that called into question the preexisting federal management system for the islands. Passage of the Marine Mammal Protection Act (MMPA) the next year triggered an increase in pressure to extend the provisions of this legislation prohibiting the killing of marine mammals to the treatment of northern fur seals.

Because the fur seal regime originated in an era dominated by the discourse of conservation with its emphasis on the achievement of maximum sustainable yields (MSY) from harvested species, it will come as no surprise that the shift toward ecosystem-based management (EBM), gathering speed during the last quarter of the twentieth century (Larkin 1977), made the fur seal regime a growing target for criticism on the part of scientists and managers, quite apart from the concerns of the environmental groups. The Bering Sea is a volatile large marine ecosystem (LME) (Sherman 1992; NRC 1996). Despite the controlled nature of the harvest, the regime became increasingly vulnerable to the proposition that arrangements dedicated to the pursuit of MSY are not appropriate, especially in a setting featuring dynamic biophysical systems and in cases where human actions are major driving forces. Conservationist arrangements still have traction in some settings largely because no one has realistic proposals for replacements (e.g., the arrangements embedded in the regime dealing with trade in endangered species created under the terms of the 1973 Convention on International Trade in Endangered Species of Wild Flora and Fauna and in the regime dealing with the conservation of polar bears under the 1973 agreement among the five Arctic states). With regard to fur seals, it is probably accurate to say that the growing awareness of the complexity of the Bering Sea ecosystem along with the rise of EBM as an influential mode of thought and the development of the protectionist premises of the MMPA pushed the fur seal regime toward a situation in which it was no longer viable, despite its earlier reputation as a successful arrangement based on principles of conservation.

There is no need to treat a regime that eventually disappears as ineffectual or unsuccessful. Some regimes are so successful that they work themselves out of a job. Perhaps more common are regimes that get swept away by shifts in prevailing discourses (e.g., the shift from MSY to EBM), with their concerns being recast within the conceptual framework of a new paradigm for addressing issues of human-environment interaction. This is a useful way to think about the fur seal regime. Because this book is about institutional change, I focus on the endogenous and exogenous factors that brought about the collapse of the fur seal regime in the 1980s. But this is not to deny the contributions this regime made to the conservation of fur seals throughout much of its life span. The emergent pattern in this case is thus a period of success followed by a buildup of stresses that eventuated in the collapse of the regime.

Facts: A Brief History of the Fur Seal Regime

The story of the twentieth century regime for northern fur seals has its roots in the latter part of the eighteenth century (see table 6.1 for a timeline), starting with the "discovery" of the Pribilof Islands by the Russians and the subsequent rush to harvest sea otters and fur seals that followed (Gay 1987). Despite the distances involved, the collection of valuable furs became the principal economic rationale for the growth of the Russian presence in the Bering Sea and the area we now know as Alaska as far south as Sitka. This industry provided the impetus for the creation of the Russian-American Company, which became the principal vehicle for the pursuit of Russian interests in North America for over half a century.

The decision of Czar Alexander II to sell Alaska to the United States in 1867 for $7.2 million has had profound consequences from that day to this for both economic and political developments in the circumpolar Arctic. But this transition did not have major impacts on the harvesting of fur seals in the short run at least. Many Americans regarded the income generated by this harvest as one of the few tangible benefits arising from what some commentators satirized as Seward's icebox or Seward's folly.[1] The U.S. government proceeded to issue twenty-year leases to private companies granting them the privilege of harvesting fur seals with few if any restrictions motivated by a concern to conserve stocks of fur seals or even to maximize sustainable yields over time.

The unregulated harvest of fur seals on land on the Pribilof Islands under American jurisdiction and on the Commander Islands under Russian jurisdiction together with the growing Canadian and Japanese practice of harvesting seals at sea took an increasing toll on fur seal stocks in the late nineteenth century. Even those adhering to a general belief that fish

Table 6.1
Fur Seal Regime Timeline

1786–1787	Russian "discovery" of the Pribilof Islands
1799	Russian-American Company formed
1867	U.S. purchase of Alaska from Russia
1890s	Modus vivendi agreements
1893	Paris Arbitration Tribunal decisions
1910	First U.S. Fur Seal Act passed
1911	North Pacific Sealing Convention signed and ratified
1912–1917	Temporary suspension of sealing
1940	Japanese withdrawal from 1911 convention
1940s	Regime in abeyance during World War II and its aftermath
1957	Interim Convention for the Conservation of Northern Fur Seals
1957	Creation of International Northern Fur Seal Commission
1966	Second U.S. Fur Seal Act implements terms of 1957 convention
1980	Extension of 1957 convention for four years
1983	U.S. Fur Seal Act Amendments
1984	Protocol to extend 1957 convention another four years negotiated
1984	U.S. decision not to ratify the protocol
1984–1985	Collapse of the northern fur seal regime

stocks—and marine living resources more generally—are inexhaustible found the evidence regarding the depletion of the fur seal herds undeniable by the 1880s and 1890s.[2] Because there was no way to solve this problem—or even to alleviate it significantly—without curtailing pelagic sealing, the United States—and, to a lesser extent, Russia—began to take actions designed to limit pelagic sealing on the part of Canadian and Japanese vessels, actions that became increasingly aggressive as the depletion of the seal herds became more severe.

At the time, the jurisdiction of coastal states over marine resources extended only to the seaward boundary of the territorial sea, three nautical miles from the coast. Due to the foraging behavior of fur seals during the summer, this arrangement was insufficient to allow the coastal states to put an end to the practice of pelagic sealing. Agreement followed on a series of largely ad hoc and often bilateral measures to ameliorate the growing conflict over the harvesting of fur seals and, in the process, to avoid the outbreak of armed conflict between sealers operating at sea and American and Russian naval vessels (Mirovitskaya, Clark, and Purver 1993). Several of these agreements involving different combinations of the four countries involved—Great Britain (representing Canada), Japan, Russia, and the United States—provided temporary relief from this problem. More generally, the United States sought to strengthen its capacity to manage fur seals in the Bering Sea by embracing the mare clausum, or closed-sea, doctrine as a basis for extending jurisdictional claims. But Great Britain, the dominant sea power at the time, objected to this unilateral attempt to extend coastal-state authority. To resolve their differences, the two countries agreed in 1891 to establish an arbitral tribunal to provide clarification regarding a number of issues focused mainly on the jurisdictional foundations of the U.S. position (Lyster 1985; Gay 1987; Mirovitskaya, Clark, and Purver 1993). Many commentators have noted this action as an important milestone in the development of international environmental law. But the Paris Arbitration Tribunal failed to solve the problem of the depletion of fur seal stocks resulting largely from pelagic sealing by the Canadians and the Japanese. The decision of the tribunal, handed down in 1893, generally supported the views of Great Britain regarding the applicable law of the sea, thereby rejecting the foundations of the U.S. claims to extended jurisdiction over fur seals in the Bering Sea.

By the beginning of the twentieth century, the problem had become acute. The seal herd on Robben Island (first under Japanese jurisdiction and later under the jurisdiction of the Soviet Union) was gone. The Commander Islands herd was down to an estimated 4,500 animals. Even the main herd on the Pribilof Islands was down to some 130,000 animals. The time had come to consider cooperative measures to prevent the extermination of northern fur seals biologically as well as commercially. Still, progress toward the development of an international regime was slow. Elihu Root, the U.S. secretary of state, proposed in 1909 the convening of a four-power conference aimed at reaching agreement on the terms of a convention establishing an international regime to conserve fur seal stocks and to pave the way toward achieving maximum sustainable yields on an ongoing basis. The eventual result, following a personal intervention on the part of the U.S. president William Howard Taft, was the emergence of agreement among Great Britain (representing Canada), Japan, Russia, and the United States on the terms of the North Pacific Sealing Convention, which was signed in July 1911 and entered into force in December of the same year (Mirovitskaya, Clark, and Purver 1993).

The convention, based squarely on the precepts of the conservation movement flourishing during the early twentieth century (Hays 1975), established a regime that was simple but ingenious. Article I of the convention prohibits all pelagic sealing in order to assure, in the words of the preamble, "the preservation and protection of the fur seals which frequent the waters of the North Pacific Ocean," thereby ensuring that all harvesting of fur seals must take place on the Pribilof and, to a lesser extent, the Commander Islands.[3] Recognizing the common-property character of wildlife, Article X of the convention calls on the United States to provide both Canada and Japan with 15 percent of the annual harvest of seal skins; Article XI commits the United States to pay $200,000 each to Great Britain (representing Canada) and Japan as compensation for their agreement to terminate all pelagic sealing. Enforcement of these arrangements was left to the governments of the signatory states, and there was no provision for a commission or any similar body to administer the provisions of this regime on a day-to-day basis.

Those who have studied the operation of this regime generally emphasize its early success (Lyster 1985). Following a temporary suspension of the harvest from 1912 to 1917, fur seal herds grew steadily, allowing harvesting to resume. This action, centered on the Pribilof herds, took the form of the culling of bachelor males on land. While the Commander Islands herds never reached commercially significant levels, the Pribilof herds reached a level of one to two million animals, thought to have approximated the carrying capacity of the Bering Sea ecosystem for this species at the time. During the 1920s and 1930s, sizable harvests took place year after year without adverse effects on the overall population.

The regime showed every sign of fulfilling the ideal that environmental historians describe as conservation and the "gospel of efficiency" (Hays 1975).

In 1940, with the onset of World War II, Japan withdrew from the fur seal regime. But the war did not affect the fur seal herds adversely. It triggered a de facto moratorium on sealing that proved beneficial for the herds, and all the parties found it in their interests in the aftermath of the war to revert to the arrangements created in the 1911 convention. In 1957, Canada, Japan, the Soviet Union, and the United States formalized this practice in the Interim Convention, an agreement replicating the terms of the 1911 convention with the addition of a North Pacific Fur Seal Commission charged, for the most part, with addressing scientific issues relevant to managing the consumptive use of fur seals. The Fur Seal Act of 1966 implemented the provisions of the Interim Convention within the United States. At this stage, the overall situation seemed stable; most observers concluded once more that the regime was a model of international cooperation based on the principles of conservation.

The period from the mid-1960s through the mid-1980s featured dramatic changes that eroded the foundations of the fur seal regime and led to its final collapse in 1985. A unique combination of forces produced this result. By the 1970s, the idea of maximum sustainable yield as the principal objective of resource management was subject to increasingly severe attacks led by scientists who were developing the alternative view we now think of as ecosystem-based management (Larkin 1977). This was also a period of rapid growth in the animal rights movement (Singer 1975), a development that fueled growing opposition to the harvesting of fur seals on ethical or moral grounds. Two institutional factors, occurring independently of these shifting views regarding the proper goal of resource management, contributed to the growth of doubt about the retention of this venerable regime. Changes in the law of the sea, formalized in the 1982 UN Convention on the Law of the Sea, extended the jurisdiction of coastal states, creating Exclusive Economic Zones extending out to two hundred nautical miles from the coast. This did not give the Soviet Union and the United States jurisdiction over fur seals throughout their life cycles. But it greatly enhanced the ability of the coastal states to eliminate pelagic sealing, regardless of the interests of distant-water states like Canada and Japan. The advent of the Reagan administration in the United States in 1981 also brought new perspectives on the proper role of the state, including a presumption against arrangements, such as the system for harvesting fur seals, amounting to state-owned and -operated enterprises. To top off this cocktail of changes, evidence began to accumulate regarding unusual volatility in the Bering Sea treated as an LME (NRC 1996). Though no evidence linked the modest annual harvest of fur seals to these changes, the size of the Pribilof Islands herds began to decline (NRC 2003).

Taken together, these developments stimulated drastic action on the part of the United States. Congress passed the Fur Seal Act Amendments of 1983 effectively terminating the Pribilof Islands Program and with it the role of the federal government in conducting the commercial harvest of fur seals. Although the United States did participate during 1984 in negotiating a protocol to the Interim Convention extending the life of the regime for an additional four years, it soon became apparent that the Senate would not vote to ratify this protocol. This unilateral action by the United States violated the spirit of the fur seal regime under which consultation was a prominent norm. But the die was cast. The rationale for the regime disappeared with the withdrawal of the United States and the assertion of U.S. control over the management of all the waters surrounding the Pribilof Islands. By 1985, this venerable regime, widely regarded in earlier times as a success story, had simply disappeared from the scene.

Analysis: Sources of Collapse

While some regimes experience declining effectiveness and even become dead letters, total collapse followed by disappearance is not common in the world of international environmental governance. It is noteworthy when total collapse does occur, especially when it affects regimes long viewed as successes in the struggle to devise effective formulas for international cooperation regarding natural resources and the environment. Although the brief history set forth in the preceding section identifies a number of factors that played a role in the demise of the fur seal regime, it is worth stopping to examine what happened in this case more systematically. The key element in the analysis I present involves a particular kind of misalignment between endogenous and exogenous factors. The fur seal regime was not designed to adapt to major changes in biophysical and socioeconomic circumstances. With the passage of time, changes in these circumstances accelerated, eventually producing a sea change in the setting in which the regime operated.

Endogenous Factors

The fur seal regime took the form of a specific deal carefully crafted to solve a problem arising in a particular time and place. Unlike fish, traditionally treated as common property resources available to all harvesters for the taking, northern fur seals came under the jurisdiction of the coastal states while resident on the rookeries of the Pribilof and Commander Islands during the summer months. Still, the coastal states were unable to control the actions of others affecting fur seals during their annual migrations. So long as territorial waters extended out only to three nautical miles, the coastal states could not regulate all harvesting even when the seals were based on their Bering Sea rookeries but moving beyond the three-mile limit during normal feeding and foraging. The deal struck in 1911 featured an ingenious solution to this unique situation. The convention prohibited pelagic sealing outright "in waters of the North Pacific Ocean north of the thirtieth parallel of north latitude and including the Seas of Bering, Kamchatka, Okhotsk, and Japan" (Art. I) and gave naval officers of all the contracting parties the authority needed to seize violators of this prohibition. The effect of the prohibition was to grant the coastal states—mainly the United States—full authority to manage the conservation and harvesting of fur seals. In return, the United States undertook to provide Canada and Japan with fixed shares of the annual harvest (Art. X) together with an "advanced payment" of $200,000 at the time the convention went into effect at the end of 1911 (Art. XI). This arrangement was well matched with the main features of the problem during the late nineteenth and early twentieth centuries. But it included no provisions to ensure resilience by adjusting the terms of the arrangement to cope with changing circumstances regarding the nature of the problem. Ingenious as it was, the regime was also rigid; it lacked the capacity to adapt to changes in the nature of the problem or in the broader setting.

Equally important, the fur seal regime was rooted firmly in a paradigm or way of thinking about resource management that gained ascendancy around the turn of the century and in the early years of the twentieth century (Hays 1975). Often associated with the ideas of Gifford Pinchot and other leaders of the Roosevelt administration in the United States, the resultant discourse emphasized the importance of conservation and the idea of scientific management, but it did so within a framework stressing the commodity value of natural resources and the desirability of harvesting living resources in a manner that would allow for sustainable yields over time. As an antidote to the destructive practices of the late nineteenth century, this doctrine of conservation and the "gospel of efficiency" was progressive (Hays 1975). But it rested on a mode of thought that was contested even at the time by influential individuals like John Muir and that proved hard to adjust over time in response to new perspectives on human-environment relations articulated in the works of seminal thinkers like Aldo Leopold (Leopold 1970). The rigidity arising from the fact that the fur seal regime was embedded in this mode of thought did not become apparent right away. For several decades, the regime performed well; the size of the seal herds grew to reach the carrying capacity of the Bering Sea ecosystem, even as a significant harvest took place on an annual basis. Yet the regime provided no means of adapting to changes in thinking about human-environment relations in the postwar years. The 1957 Interim Convention did add a commission (Art. V). But the convention specifies that the commission is to focus on scientific research and analysis needed to prevent the herds from falling "below the level of maximum sustainable productivity" (Art. V). When new ideas about human-environment relations began to take root and to gain influence in the 1950s and 1960s, the fur seal regime was held hostage by a mode of thought whose grip on thinking about policy regarding resource management was on the wane.

The fur seal regime's approach to decision making served to reinforce these endogenous sources of rigidity. The 1911 convention enjoined each of the contracting parties individually to "enact and enforce such legislation as may be necessary to make effective" the provisions of the regime (Art. VI), and it called upon the parties to "cooperate with each other in taking such measures as may be appropriate and available" to prevent pelagic sealing (Art. VIII). But it made no provision for joint decision making to address emerging issues. The result was a set of relatively strong measures to implement the deal enshrined in the provisions of the convention coupled with an assumption that what we think of now as adaptive management or governance would not be a major concern for this arrangement (Lee 1993). The 1957 Interim Convention took one small step forward from the provisions of the 1911 convention in this realm. While creating a commission, this convention also specified that "[e]ach Party shall have one vote" and required unanimity in order to approve any decision or recommendation (Art. V[4]) relating to the management of fur seals. Bearing in mind that the role of the commission was confined largely to scientific matters, it is fair to conclude that even the postwar version of the fur seal regime did not anticipate the need to provide for adaptive governance in any significant way. The conclusion seems inescapable that this regime was designed to operate effectively only in a stable biophysical and socioeconomic setting.

Turning to the issue of provisions for review and amendment procedures, the story is much the same. Although the 1911 convention did not include an amendment procedure as such, it did allow any of the parties to call for a conference to "consider and if possible agree upon . . . such additions and modifications, if any, as may be found desirable." No such conference was held during the life span of the 1911 convention. The Interim Convention, too, lacked explicit procedures governing amendments. But it did call upon the parties to "meet in the twenty-second year after entry into force of the Convention . . . to determine what further agreements may be desirable in order to achieve the maximum sustainable productivity of the North Pacific fur seal herds" (Art. XI). Article XIII contains an even more encompassing commitment, stating that at the request of any party, "representatives of the Parties will meet at a mutually convenient time within ninety days of such request to consider the desirability of modifications of the Convention." Any resultant modifications would require unanimous agreement on the part of the regime's four members. But there is no indication that any party ever called for the use of these procedures under the terms of either the 1911 convention or the 1957 Interim Convention.

Article XVI of the 1911 convention specified that the convention—signed on July 7, 1911—was to go into effect on December 15 of the same year. It stated that the convention would remain in force for fifteen years from December 15, 1911, and thereafter until twelve months following written notice of withdrawal on the part of any one of the parties. The provisions of the 1957 Interim Convention were somewhat more advanced. Article XIII(4) stated that the convention would remain in force for twenty-two years and "thereafter until the entry into force of a new or revised fur seal convention between the Parties, or until the expiration of one year after such period of twenty-two years, whichever may be the earlier." The 1911 convention lapsed following the withdrawal of Japan in 1940. The circumstances surrounding the demise of the 1957 Interim Convention are more complex. The parties signed a protocol amending the Interim Convention in 1980, which provided for, among other things, an extension of the life of the convention. By 1984, circumstances had changed. The U.S. Congress had passed the Marine Mammal Protection Act of 1972 and the Fur Seal Act Amendments of 1983, and several players had begun to express criticisms of the regime on a number of grounds. The parties did negotiate and sign a protocol on October 12,

1984, extending the life of the Interim Convention but calling for a review of the operation of the regime within two years after the entry into force of the protocol. But by then, it was too late. It had become apparent that the U.S. Senate would not ratify such a protocol, leaving the Interim Convention itself without a legal basis for continuing in force. The notes accompanying the recorded text of the Interim Convention indicate that it expired on October 14, 1984. The juxtaposition of the two dates—October 12 and 14—seems odd in this context. But the basic message is clear. The fur seal regime in its familiar form was no longer viable, and it lacked any capacity to engage in what we now call adaptive governance. Collapse became inevitable.

The fur seal regime took the form of a solution to a specific problem that worked well at first but lacked the capacity to adapt as the nature of the problem—or perceptions of the nature of the problem—evolved. This is not to take away from the roles played by key individuals, including the British foreign secretary Lord Salisbury in driving the negotiations of the 1890s and U.S. president Taft and naturalist David Starr Jordan in calling attention to the issue and brokering the deal encapsulated in the terms of the 1911 convention (Mirovitskaya, Clark, and Purver 1993). As the next subsection details, however, circumstances changed at an accelerating pace in the late twentieth century. What proved successful in the conditions prevailing in the 1910s through 1930s became increasingly outmoded in the decades following the end of World War II. The regime itself had no capacity to adapt or adjust in such a way as to remain well matched with major features of the environment in which it operated.

Exogenous Factors

The principal message arising from this case regarding exogenous factors is straightforward. The world changed around the fur seal regime in ways that produced a growing mismatch between the regime and the broader setting in which it operated. That much is clear. Still, it is worth taking the time to consider these changes in some detail. One way to characterize the story in this case is to think in terms of multiple interactive stresses that overwhelmed the robustness of the regime—its capacity to operate effectively without making significant changes in its principal elements (Anderies, Janssen, and Ostrom 2004). The regime lacked resilience in the sense of a capacity to make significant adjustments to changing circumstances without compromising its defining features (Holling and Gunderson 2002).

Wars played an important—and oddly positive—role in the life of the fur seal regime. Little commercial sealing occurred during the years of World War I (1914–1918) and World War II (1939–1945). The result was a de facto moratorium on sealing during these periods that played some role in helping the seal herds to recover and to remain healthy in biophysical terms. As astute observers have pointed out, the turbulent history of the twentieth century would have made the creation of an effective fur seal regime impossible during much of the rest of the century (Mirovitskaya, Clark, and Purver 1993). Within a few years after the successful conclusion of the 1911 convention, for instance, the onset of World War I would have precluded any effort to establish a regime dedicated to the conservation of fur seals. The parties reached agreement on the convention during a rare window of opportunity. But with the convention in place, the war became an exogenous force that produced results beneficial to the conservation of fur seals.

By contrast, changes in surrounding institutional arrangements, at the domestic and international levels, proved disruptive to the fur seal regime (Young 1987). In the United States, a series of laws brought about dramatic changes in the legal environment in which the regime operated. The Alaska Native Claims Settlement Act of 1971 (ANCSA) created village corporations for the communities of St. Paul and St. George on the Pribilof Islands and turned over title to most of the land on the islands to these corporations. The Marine Mammal Protection Act of 1972 introduced a new regime for the management of marine mammals in areas under the jurisdiction of the United States, featuring a general ban on intentional killing of individual animals. The Fishery Conservation and Management Act of 1976 established a two-hundred mile Fishery Conservation Zone adjacent to coastal areas of the United States and asserted authority on the part of the federal government to manage activities involving living resources in this zone. The Fur Seal Act Amendments of 1983 put an end to commercial sealing as a state-owned and -operated industry. Taken together, these domestic U.S. legislative initiatives produced radical changes in the socioeconomic setting of the fur seal regime. The United States asserted extended authority to ban pelagic sealing, withdrew from its prior role in the harvesting of fur seals, and created an alternative approach to the management of marine mammals based on a policy of avoiding intentional killing of these animals. There is no way the regime established under the terms of the 1911 convention and the 1957 Interim Convention could have survived these changes without a far greater capacity to adapt to changing circumstances than it possessed.

Parallel changes at the international level served to reinforce and intensify the impact of these domestic institutional changes. The most important development here was final agreement on the terms of the 1982 UN Convention on the Law of the Sea. A number of provisions of this convention served to increase the effective control of coastal states over marine resources. By far the most important of these are the provisions set forth in Part V of UNCLOS on the Exclusive Economic Zone (EEZ). Extending two hundred nautical miles from the coast, EEZs provide coastal states with a broad grant of authority to make decisions regarding both the use and the conservation of marine living resources. The EEZs of Russia and the United States encompass the bulk of the Bering Sea as well as sizable swaths of coastal waters along the Pacific coasts of North America and Asia. Although the convention did not enter into force until 1994 and the United States has yet to ratify it, all the states concerned with northern fur seals proceeded more or less immediately to adopt Part V of UNCLOS as a matter of customary law and accepted practice. This did not make international cooperation regarding fur seals entirely irrelevant. Individual seals sometimes move beyond the boundaries of EEZs, and the rise of the idea of ecosystem-based management has heightened interest in the effects of changes in the overall Bering Sea ecosystem on individual species like the northern fur seal. But changes in the law of the sea combined with changes in domestic law to reduce greatly the incentives for the United States to maintain the fur seal regime, at least in the form established under the terms of the 1957 Interim Convention. By the mid-1980s, the legal environment was clearly adverse to the survival of this regime.

A series of biophysical and socioeconomic changes in the 1960s and 1970s added weight to these legal and institutional factors, taking a further toll on the resilience of the fur seal regime. As the 1960s gave way to the 1970s and then the 1980s, signs of volatility in the Bering Sea ecosystem began to mount (NRC 1996; Young 2005). The population of Steller sea lions declined, gradually at first but then at an increasing rate. The number of Red-legged Kittiwakes, a species endemic to the Pribilof Islands, began to fall. Fur seals themselves, whose growing numbers had led managers to engage in a deliberate herd reduction program in the 1950s, experienced a decline in numbers. There is no basis for assuming the existence of a connection between the commercial harvest and the decline in the fur seal population during this period. What has become apparent with the passage of time, however, is that the Bering Sea ecosystem as a whole is highly volatile; it responds to combinations of forces that are poorly understood (NRC 1996; NRC 2003). Scientists have advanced a variety of hypotheses about the sources of dramatic changes in the status of marine organisms of the Bering Sea, and it is likely that a number of factors operating simultaneously contribute to these developments. Nonetheless, the dynamics of this ecosystem are poorly understood. Many of those responsible for management reasoned that it would be prudent, under the circumstances, to terminate the commercial harvest of fur seals as a precautionary measure.

Changes in the economics of sealing reinforced this view (Young 1981b). At the time of the purchase of Alaska from Russia, the commercial harvest of fur seals struck many commentators as one of the few benefits arising from the acquisition of this vast and remote territory. Even as a state-owned and -operated enterprise from 1910 onward, commercial sealing was profitable. Until 1969, receipts from the sale of seal skins covered not only the cost of the harvest as such but also various forms of assistance that the federal government provided through its Pribilof Islands Program to the communities of St. Paul and St. George. Under the terms of the Alaska Statehood Act of 1958, the federal government agreed to turn over any surplus remaining after covering the cost of running the Pribilof Islands Program to the state of Alaska, and the state benefited from this arrangement throughout the 1960s. But an examination of the economics of commercial sealing reveals several other pertinent points. Whatever its economic significance in the late nineteenth century, the industry had become economically insignificant by the 1970s with the onset of large-scale oil development in Alaska and, more generally, the shift to modern industrialized systems in which activities involving the harvesting of wildlife were marginalized or transformed into recreational activities. Because commercial sealing was a state-owned enterprise, there was a dearth of influential private interests motivated to lobby for the continuation of this industry. Those whose livelihoods and lives were affected—the Aleut residents of the communities of St. Paul and St. George—were not in a position to bring effective pressure to bear on agents of the federal government. As the 1970s turned into the 1980s, therefore, the economic basis for the continuation of a commercial harvest of fur seals evaporated, taking with it the rationale for the sort of regime established under the terms of the 1957 Interim Convention.

While the economic significance of the harvest of fur seals waned, environmentalism was on the rise as a force to be reckoned with, especially during the period following the publication of Rachel Carson's book *Silent Spring* in 1962 and the emergence of the preservationist movement during the 1970s. The bifurcation of the environmental movement into a mainstream that was increasingly concerned about the impacts of human actions on complex ecosystems and a sizable sidestream that focused more on the rights of animals was problematic for some purposes. But both streams provided support for those opposed to the governance approach embedded in the fur seal regime. Whether opponents of the continuation of commercial sealing started from the need to introduce ecosystem-based management as an alternative to efforts to achieve maximum sustainable yields of individual species or from a desire to recognize the rights of animals, they had no difficulty finding common ground in opposing the form of sealing called for under the terms of the fur seal regime. The 1970s marked a watershed in these terms, at least in the United States. What had been for the most part a protest against the orthodox view of human-environment relations became a force to be reckoned with capable of stimulating the passage of legislation and altering human preferences on a large scale (Hays 1987). The fur seal regime was not a major focus of the larger debate about these matters. But it is not coincidental that the demise of the regime came in the wake of these developments.

Even the fashion industry played a role in the array of exogenous forces coalescing to call into question the rationale underlying the regime. Whereas wearing fur had once been stylish, leading designers turned to other materials in the 1960s and 1970s. Whether the campaigns of the animal-rights movement played a role in this shift or it is better interpreted as a response to other factors is debatable. But the fact remains that the high-end market for seal skins declined during the period. Without this market, the value of seal skins, including those of harp seals as well as fur seals, plummeted (Wenzel 1991). All things considered, it was hard to find anyone able and willing to provide strong support for the continuation of commercial sealing by the 1980s.

Endogenous-Exogenous Alignment

The provisions of the fur seal regime produced an arrangement well aligned with the early-twentieth-century setting in which it operated. The prevailing law of the sea did not grant coastal states sufficient authority to control the harvest of seals by banning pelagic sealing. The four states involved—Great Britain (representing Canada), Japan, Russia, and the United States—were all major powers; there was no prospect of a dominant actor or hegemon emerging in this issue area able to impose a regime on the others. The economics of commercial sealing provided a number of actors with strong incentives to harvest seals under a variety of pretexts. The basic deal embedded in the 1911 convention—coastal state control of the harvest with a fixed proportion of the seal skins to go to the distant-water states—made sense, and the parties agreed to deploy their navies to deter or prevent would-be violators from circumventing the terms of the agreement. The fur seal herds—especially those based on the Pribilof Islands—made a strong comeback in the years following the creation of this regime. As we now know, the Bering Sea ecosystem is complex and dynamic; a variety of factors undoubtedly played some role in the revival of the seal herds between the 1910s and the 1930s. But virtually everyone who has examined this case with care has concluded that the regime played an important role in the renewal of the herds (Lyster 1985). For many years, commentators routinely pointed to this regime as an exemplar of success in achieving the protection of wildlife through international cooperation.

As the preceding subsections make clear, the gap between the basic character of the fur seal regime and a number of exogenous factors widened with the passage of time. This development involved a convergence of institutional, biophysical, and socioeconomic factors. The broader legal setting changed, granting coastal states considerably greater authority over marine living resources than they enjoyed in 1911. Awareness of the dynamism of the Bering Sea ecosystem increased—and at an accelerating rate—with the passage of time. While the significance of commercial sealing declined over time, the influence of the environmental movement—including its animal-rights wing—grew. Taken together, these factors gave rise to a classic case of cumulative stress calling into question not only the robustness but also the resilience of a regime whose creators had experienced no incentive to pay attention to them.

The resultant pressures might have overwhelmed any regime designed to manage the consumptive use of fur seals. But the particular regime established under the terms of the 1911 convention and reproduced largely unchanged under the terms of the 1957 Interim Convention lacked flexibility. The Interim Convention does refer to the possibility of meeting "to consider the desirability of modifications of the Convention" (Art. XIII[4]). But there are no provisions for altering the regime's defining feature as an agreement motivated by the idea of conserving an individual species in order to achieve maximum sustainable yields from an ongoing harvest. Any changes, including those limited to modest modifications, would have required unanimous agreement among the parties. Even the 1957 version of the regime, with its provision for a North Pacific Fur Seal

Commission, lacked any mechanism for considering institutional innovations or debating the relative merits of alternative approaches to restructuring the regime to meet changing exogenous conditions. During the final phase of the regime, the resultant misalignment became dramatic, a fact that helps to explain why the fur seal regime simply collapsed and vanished without a trace, when the end came in 1985.

There is no reason to write off this case as an example of failure. The fur seal regime addressed a problem that was acute during the early years of the twentieth century and created a governance system that performed well over a considerable period. But with the passage of time and the onset of major changes in exogenous factors, the regime proved unable to make the adjustments needed to remain viable. Similar stories are common at all levels of social organization, especially in cases like this one in which the relevant changes encompass fundamental shifts in the discourses or paradigms (e.g., the shift from MSY to EBM) key players rely on to conceptualize the nature of the problem to be solved. The failure of governance systems to adapt under these conditions should not detract from an acknowledgment of their achievements during earlier phases. Perhaps it would help make this point clear to label the emergent pattern in this case early success followed by eventual collapse.

Forecast: The Road Ahead

There is no road ahead for the fur seal regime as such. The collapse of the regime in 1984 and 1985 was decisive. The age of managing the harvest of individual species in pursuit of maximum sustainable yields is over. No one now mourns the demise of a regime rooted in circumstances that prevailed a century ago. No actor interested in the living resources of the Bering Sea region will propose the creation of a new regime focusing on northern fur seals as an individual species. This is the end of the story regarding this case study. But the problem of managing human actions affecting the fur seals of the Bering Sea region has not gone away. It has simply become one element in the larger problem of managing human-environment interactions in the region as a whole.

What options are available to those endeavoring to make progress in building a new and more comprehensive regime for the Bering Sea region (Young 2005)? There is no simple answer to this question. Yet some observations about the limitations of familiar approaches will help to focus our thinking about the issue. The standard preservationist arguments are of limited value in addressing this issue. With the exception of the whales and walrus taken by indigenous peoples in the Bering, Beaufort, and Chukchi Seas, no marine mammals have been taken intentionally in the Bering Sea region for twenty-five years. Yet populations of sea otters, sea lions, and fur seals continue to decline for reasons that are complex and not well understood. Partly, this is a result of by-catch in the large commercial fisheries of the region or entanglement with ghost nets lost or discarded by commercial fishers. In part, it is a consequence of complex processes occurring in the Bering Sea region as a highly dynamic socio-ecological system (see table 6.2). We do not understand the behavior of this system well enough to act on the precepts of preservationism. Even a cessation of commercial fishing—an action that has no traction in terms of policy—would offer no guarantee of recovery of the stocks of the relevant marine mammals.

For those who find normative arguments appealing, a more attractive option might be an application of Aldo Leopold's concept of "biotic citizenship" to the Bering Sea region (Leopold 1970). Leopold was no opponent of consumptive uses of living resources. But during his later years he developed an ethical perspective on human-environment relations based on the proposition that humans should act as biotic citizens, as responsible members of complex socioecological systems rather than as outside conquerors with a license to impose their will on biophysical systems, regardless of the broader and longer-term impacts of their actions. An application of this idea to an area like the Bering Sea region would require a massive effort to apply the concept of biotic citizenship to real-world conditions. But the idea itself has much to offer to those interested in introducing a new discourse relating to human-environment interactions.

A more realistic approach may be to bring the precepts of ecosystem-based management to bear on this issue area (Norse and Crowder 2005). As a paradigm for thinking about human-environment interactions, EBM has replaced the doctrine of maximum sustainable yield among resource managers in many quarters. There is even experience with the use of this new frame of reference at the international level. The regime created under the terms of the 1980 Convention on the Conservation of Antarctic Marine Living Resources (CCAMLR), for instance, calls explicitly for the use of an ecosystem approach. Nevertheless, the move to EBM as a basis for managing human-environment interactions in a setting like the Bering Sea is easier said than done. Several ways forward are possible, but none offers a simple solution to the problem at hand.

Table 6.2

Potential Causes of Declines in Abundance of Fur Seals, Sea Lions, and Harbor Seals in the Eastern Bering Sea and Gulf of Alaska

	Likelihood of involvement in declines since 1980			
Cause	Fur seal	Sea lion	Harbor seal	Comment
Disease	Low	Low	Low	Few data, but no evidence of increased effects that could explain declines
Chemical pollution	Low	Low	Low	Few data; levels low relative to known effects in other populations
Entanglement	Moderate	Low	Low	Could have been only a contributing factor in the fur seal declines
Harassment	Low	Low	Low	Local effects only; not important for the geographic scale over which the declines have occurred
Commercial harvest	Low	Low	Low	Could not explain observed declines beyond the mid-1970s
Subsistence harvest	Low	Low	Low	Not a factor in the widespread declines
Incidental take	Moderate	Moderate	Low	Contributed to the declines, but not a major factor; cannot explain continued declines
Predation	Low	Low	Low	Few data; could not explain the declines, but may affect recoveries
Direct climate effects on survival	Low	Low	Low	Cannot explain the declines
Fishery effects on food availability	Moderate	High	High	Could have played a major role in all declines
Climate effects on food availability	Moderate	Moderate	Moderate	Could have played a role in all declines
Competition from fish predators	Moderate	Moderate	Moderate	Could be a major factor in the declines, given observed changes in community structure that were most likely caused by commercial fishing and environmental changes

Source: The Bering Sea Ecosystem. Committee on the Bering Sea Ecosystem; Polar Research Board; Commission on the Geosciences, Environment, and Resources; National Research Council. Washington, DC: National Academy Press, 1996, http://www.nap.edu/openbook.php?record_id=5039&page=145.

One approach emphasizes ecosystem-based fisheries management (EBFM), the development of fisheries management plans that take into account the broader biophysical setting in which targeted species are located (Pikitch et al. 2004). The North Pacific Fisheries Management Council, the U.S. body responsible for managing the commercial fisheries in the sector of the Bering Sea region under U.S. jurisdiction, has become a leader in this movement (Fluharty 2005). But this broadening of the perspective of fisheries management is hardly a solution to the problem under consideration here. The focus remains a matter of maximizing the productivity of the fisheries subject to certain biophysical constraints. There are no clear-cut criteria regarding how to integrate the provisions of the Fishery Conservation and Management Act with those of other relevant statutes such as the Marine Mammal Protection Act. Above all, this approach does not offer a constructive way of dealing with issues, like the fate of fur seals, that are not confined to the areas under the jurisdiction of individual coastal states and that therefore call for a concerted effort at the international level. Those concerned with the fate of the northern fur seal are unlikely to settle for an application of EBFM in this realm.

Another approach recognizes the multiplicity of factors that must be considered in any full-blown effort to practice EBM and takes the view that to be successful, EBM must concentrate on interactive forces operating in well-defined and somewhat limited spatial domains (Young et al. 2007). The idea here is to address problems of fragmentation and of spatial and temporal mismatches by instituting intensive management practices in delimited areas or places, introducing systems featuring ocean zoning, and creating consultative arrangements that require all relevant management agencies to engage in meaningful collaborative processes (Crowder et al. 2006). This place-based option is well suited to some situations. It has produced major gains in efforts to manage areas like the Australian Great Barrier Reef (Day 2002). But it is difficult to see how this approach can help northern fur seals who migrate east and west within the Bering Sea in considerable numbers and who move far beyond the confines of the Bering Sea region during their migratory cycles. At a minimum, place-based management would require an ambitious form of multilevel governance extending from local communities like St. Paul and St. George all the way up to the international level encompassing Canada and Japan as well as Russia and the United States.

A currently popular approach to EBM directs attention to what are now widely known as large marine ecosystems (Sherman 1992). The emphasis here would focus on the advantages of considering the Bering Sea ecosystem as a whole. This way of thinking has intuitive appeal. But it also has serious limitations from the point of view of those concerned with the fate of seals following the demise of the fur seal regime. Not only does the annual migration carry seals beyond the confines of the Bering Sea region; our understanding of the Bering Sea region as a biophysically volatile LME also makes it hard to devise policies that provide any assurance regarding the status of the fur seal herds and that are likely to prove acceptable to representatives of major interest groups. For all its intuitive appeal, the LME concept has proven more effective as a declaratory ideal than as an approach to management capable of achieving desired results at the operational level.

There is no bright line to guide the thinking of those concerned with the welfare of fur seals or, for that matter, other major stocks of marine mammals in the North Pacific. The old regime is dead; there is no prospect of reviving this outmoded approach to governing human-environment interactions. But it is far from clear how to move beyond the old species-specific arrangements in such a way as to provide reasonable assurance that the fur seal herds will survive and even flourish. The best we can say at this juncture is that there is general recognition of the importance of the resultant challenge. Any workable solution that emerges in a specific area like the Bering Sea region will be of great interest to resource managers and environmental advocates all over the world.

Conclusion

The fur seal regime arose to address an acute problem and remained in place over a number of decades, even bouncing back intact after the severe disruption of World War II. For some time, the fur seal herds grew under the stewardship of this regime, supporting a commercial harvest that was profitable until 1969. Most of those who have studied the regime have emphasized its success over a relatively long period of time. A good many have treated the regime as an exemplar of success, at least among arrangements dedicated to conserving and maximizing the productivity of individual species.

Yet the regime collapsed and sank without a trace in the 1980s. The interesting story centers on the alignment (or misalignment) of endogenous and exogenous forces during the regime's lifetime. The regime created in 1911 provided an ingenious solution to a rather specific problem. But neither the 1911 treaty nor its successor, the 1957 Interim

Convention, included effective measures for adapting or adjusting the regime in the face of biophysical or socioeconomic changes. While this was not a serious defect in the early years, the regime's rigidity emerged as a problem when circumstances changed, slowly at first but then more dramatically during the 1960s and 1970s. By the time the regime collapsed in 1985, the misalignment was profound. The regime was outdated and outmoded; it rested on a view of the world and a paradigm for thinking about human-environment interactions that had crumbled in the face of major changes in the economic, legal, organizational, and political circumstances. The cumulative stresses engendered by these changes pushed the regime past a tipping point where it simply disintegrated in much the same manner as other notable institutional collapses (e.g., the collapse of the Soviet Union) occurring at the domestic level as well as the international level. What began as a success story ended in collapse.

What happens next regarding the fate of the fur seal herds of the Bering Sea? This is an important question, but one that is difficult to answer in any convincing way at this time. There is no prospect of a return to governance systems focused on individual species and designed to maximize productivity in a context of consumptive uses. Yet the alternatives are neither fully developed analytically nor broadly accepted in the world of policy. Perhaps the best bet in these terms is some form of ecosystem-based management that treats fur seals as one component of a larger ecosystem and that recognizes the complexity and volatility of the Bering Sea ecosystem. But we are still a long way from such an approach that can provide an operational procedure for making management decisions pertaining to fur seals along with other components of this ecosystem. Developing this approach to a point where it is able to cope with matters of this sort in operational terms remains a major challenge for resource management in the twenty-first century.

7

Toward a Theory of Institutional Change: Accomplishments and Challenges

Introduction

We have made progress in seeking to understand emergent patterns in international environmental governance. But the study of institutional dynamics remains an infant industry among those interested in environmental issues. The preceding chapters not only illuminate some aspects of institutional change but also direct attention to a range of issues that deserve attention in future research on institutional dynamics. In this concluding chapter, I take stock of the state of play in this field, endeavoring to evaluate accomplishments so far and to identify challenges for the next phase of research in this area.

I begin by analyzing the robustness of the endogenous-exogenous alignment thesis that has served as the focus of attention in analyzing the cases. I then proceed to ask whether there are identifiable patterns of institutional change above and beyond the five emergent patterns examined in the preceding chapters. This leads to a discussion of the broader settings in which environmental regimes operate and the question of whether there are underlying or deeper causal forces that form a substrate beneath the proximate drivers operative in the case studies. A discussion of cutting-edge questions and next steps in the analysis of institutional dynamics constitutes the focus of the penultimate section of the chapter. I conclude by turning to an assessment of the implications of this study of institutional dynamics for environmental policy and more specifically for the design of regimes needed to address a number of increasingly urgent problems associated with human-environment interactions in an era of human-dominated ecosystems (Vitousek et al. 1997; Steffen et al. 2004) or what many have come to regard as the Anthropocene (Crutzen 2002; Steffen, Crutzen, and McNeill 2007).

Thesis Robustness

Is the endogenous-exogenous alignment thesis robust? There are obvious limits to our ability to arrive at a clear and decisive answer to this question. The thesis does not take the form of a conventional hypothesis stating a well-defined relationship between some straightforward dependent variable and one or more independent variables. We are not now and may never be in a position to provide a quantitative measure of alignment treated as a matter of the extent to which the institutional attributes of regimes are well matched to the broader biophysical and socioeconomic settings in which they operate (Young 2002; Galaz et al. 2008). Nonetheless, the case studies do shed light on the nature of this relationship. They provide a substantial measure of support for the proposition that the alignment between the character of a regime and the nature of the setting in which it operates is a key determinant of the pattern of change that the regime exhibits over time.

The case studies illustrate several distinct forms of endogenous-exogenous alignment as well as their consequences for institutional dynamics. In the case of the ozone regime, science has produced growing evidence of the links between the production and the consumption of several families of chemicals and the extent and severity of seasonal thinning of the stratospheric ozone layer. Industry has found it relatively easy to devise affordable substitutes for ozone-depleting substances (ODSs). The regime has proven well matched to these features of the setting. It is easy to accelerate phaseout schedules for chemicals that are already regulated under the provisions of the 1987 Montreal Protocol. The regime prohibits the use of exceptions as an easy way out for individual members unhappy with specific decisions of the COP and MOP. It provides financial assistance in the form of the Multilateral Fund to assist developing countries in finding the means to comply with phaseout decisions of the COP and MOP. It offers a somewhat more complex but still workable procedure for adding new families of chemicals to the roster of regulated or banned substances once evidence of their ozone-depleting nature becomes unambiguous. The emergent pattern I have labeled progressive development is easy to account for under these conditions.

The Antarctic Treaty System (ATS) provides additional insights regarding the endogenous-exogenous alignment thesis. Antarctica is a case in which the prevailing regime has experienced a sequence of more or less severe stresses or challenges emerging from the broader setting in which it operates. Arising initially from a desire on the part of developing countries to share in the proceeds from resource development in the region, and moving forward to growing concerns about environmental impacts of activities like commercial fishing and rising pressure from the environmental community to turn the entire region into a wilderness area, the challenges to the regime have generated serious threats to the resilience of the ATS. But the ATS has exhibited a remarkable capacity to rise to these challenges and to find ways to alleviate the resultant stresses, while retaining its defining commitments to demilitarization, the freezing of jurisdictional claims, and the protection of the ecosystems of the south polar region. It has evolved into a regime complex adding components that have played key roles in meeting challenges as they arise. While decisions on substantive issues require unanimous support on the part of the consultative parties, the number of parties in this category is relatively small and the parties have clear incentives to avoid the likely consequences of deadlock. The result is an emergent pattern that I have labeled punctuated equilibrium, a pattern marked by oscillation between challenge and response in a manner leading to the survival and even strengthening of this governance system over time.

The case of the regime for northern fur seals illustrates the impacts of sharp changes in the biophysical, socioeconomic, and cognitive settings in which regimes operate and the effects of such changes on the effectiveness of governance systems over time. From the inception of this regime in 1911 to its suspension caused by the onset of war at the end of the 1930s, this relatively inflexible arrangement operated in a benign and relatively stable setting (Lyster 1985). The regime had little capacity to adapt to changing circumstances. But this did not cause serious problems. The idea of managing consumptive uses of resources in such a way as to achieve maximum sustainable yields constituted the dominant discourse throughout this period. The fur seal population recovered nicely from a state of severe depletion prior to the creation of the regime, and the Bering Sea ecosystem appeared to be stable. All this changed in the decades following World War II. MSY lost its status as a dominant discourse. Attitudes toward consumptive uses of marine mammals shifted, slowly at first but more rapidly after crossing a threshold to achieve mainstream status. It became clear that the Bering Sea ecosystem is marked by dynamic processes that are poorly understood (NRC 1996). The fur seal regime, whose inflexibility had not been a source of problems in earlier times, proved incapable of adapting to stresses attributable to a changing setting. The endogenous-exogenous alignment thesis provides a clear explanation of what happened in this case. In the early decades, a simple and rather rigid regime operated effectively in a stable setting. Later on, rigidity became an increasingly severe limitation. The regime was unable to adapt to shifting circumstances in the Bering Sea ecosystem and in the dominant discourse applicable to issues featuring human-environment interactions relating to marine mammals.

The remaining cases—climate change and whaling—also lend credence to the endogenous-exogenous alignment thesis. The climate regime, at least in its current form based on the UNFCCC and the 1997 Kyoto Protocol, is not up to the task of addressing the problem of climate change. The gap between the regime itself and the setting in which it operates has widened with the growing awareness of the scope and scale of the impacts of the problem. We can now say with considerable confidence that the impacts of climate change will be severe, and we have come to understand that the Earth's climate system could produce abrupt changes that will stretch existing adaptive capacity to the limit. But the regime is limited in its current form as an arrangement featuring politically determined targets and timetables, and it is cumbersome to adjust to changes in our understanding of the problem of climate change. Neither the United States nor China, which together account for 35 to 40 percent of greenhouse gas emissions, has accepted any binding commitments to reduce emissions under the current arrangements; full compliance with the terms of the Kyoto Protocol would hardly make a dent in the problem as we now understand it, and current efforts to reform the regime seem extremely modest given our emerging understanding of climate change and its consequences. This does not rule out the prospect of crossing a threshold leading to a watershed in the character of the regime in the foreseeable future. But it does justify the characterization of this regime in its current form as a case of arrested development.

The case of the regime for whales and whaling exemplifies yet another pattern of institutional change. A combination of developments featuring the decline of the whaling industry, the rise of environmentalism, and shifts in the membership of the regime eventuated in the formation of a coalition that was able to muster the three-fourths majority needed to adopt the moratorium in 1982. Since then, the regime itself has become the locus of a protracted and inconclusive battle between a coalition of antiwhalers and a looser grouping of members supporting a resumption of whaling under the terms of the Revised Management Procedure. Neither side has been able to triumph in this battle. The moratorium is still in place. But whaling nations like Japan and Norway have made use of loopholes to harvest whales on a significant scale. The outcome with regard to institutional dynamics is a situation in which an inflexible regime is encountering a broader setting that is turbulent and shifting in ways that are hard to anticipate, much less to steer effectively. The regime has become a battleground among stakeholders who employ radically different discourses to express their preferences. This combination of diversion followed by gridlock has marginalized the regime as a tool for governing human-environment relations affecting the fate of marine mammals (Friedheim 2001a).

The case studies make it clear that a number of the determinants of emergent patterns in environmental governance come into play in actual cases (see chapter 1, table 1.1). With regard to endogenous factors, decision rules and procedures that allow members to file reservations or objections are prominent. Given the volatility of many international problems, monitoring, reporting, and verification coupled with flexibility in the face of changing circumstances are key attributes. So also is the availability of funding mechanisms that allow developing countries to participate effectively.

Turning to exogenous factors, the familiar triad of power, interests, and knowledge stands out. The exercise of structural power on the part of the United States is an important element of the ozone story. Diverging interests between the Annex 1 countries and the G77 plus China loom large in the case of climate change. Paradigmatic shifts away from MSY and toward various forms of EBM played major roles in the cases of whales and seals. But technological changes, societal shifts, and trends in biophysical conditions also come into play as important features of specific cases. Technological advances (e.g., the introduction of the harpoon gun in commercial whaling or the high-endurance stern trawler in commercial fisheries) can overwhelm regimes that have worked relatively well in prior periods. New actors, including nonstate actors like the Antarctic and Southern Ocean Coalition in the case of Antarctica and the Intergovernmental Panel on Climate Change in the case of climate change, can have major impacts on emergent patterns in environmental governance. Biophysical changes (e.g., dramatic changes in the Bering Sea ecosystem during the second half of the twentieth century) can undermine the fit between regimes and problems, especially in cases where the relevant regimes are rigid or difficult to adjust. The take-home message here is simple: Numerous endogenous and exogenous factors are important in specific cases. But it is the alignment between these two categories of factors rather than the character of individual factors that determines the outcome with respect to emergent patterns in actual cases.

This account draws on observations and insights emerging from in-depth studies of real-world cases. Additional insights come into focus when we approach the relationship between internal and external forces in a more analytic and necessarily simplified manner. Dichotomizing both the internal and the external forces at play yields a two-by-two matrix in which the regime itself is either rigid/inflexible or adaptable/flexible and the broader setting is either stable or volatile (see table 7.1). An examination of the individual cells of this matrix points to several combinations of endogenous and exogenous factors that occur regularly in the world of international environmental regimes, engendering distinct types of institutional dynamics.

The upper-left-hand cell of the matrix encompasses cases in which more or less rigid regimes operate in stable settings. So long as biophysical conditions remain unchanged, the number and identities of the relevant players stay the same, and there are no radical shifts in prevailing discourses, institutional rigidity need not present a problem. Many efforts to create environmental governance systems assume, at least implicitly, that real-world situations fall into this category, a condition I regard as success within limits. As the initial incarnation of the regime for northern fur seals demonstrates, circumstances of this kind may prevail for lengthy periods without running into stresses that are severe enough to precipitate significant institutional changes.

Such arrangements are ill equipped to thrive or even to survive, however, in volatile settings featuring changes in prevailing discourses (e.g., the transition from MSY to EBM), the emergence of new players or shifting alliances among current members, or sharp shifts in the condition of key biophysical systems (e.g., state changes in physical systems brought about by conditions like climate change). This is the situation we face today in a number of issue areas. The climate regime, at least in its current form, shows little capacity to adjust in response to the emerging sci-

Table 7.1
Endogenous/Exogenous Alignment

	SETTING	
REGIME	Stable	Volatile
Rigid	Fur seals I	Climate Fur seals II
Flexible		Ozone ATS

entific consensus regarding the tolerance of the climate system for rising concentrations of greenhouse gases (GHGs) and the growing awareness of the possibility of nonlinear and abrupt changes in this system (Pearse 2007). The regime designed to combat desertification lacks the capacity to address the complex mix of biophysical and socioeconomic forces that account for the progressive degradation of land not only in the developing world (e.g., the arid regions of Africa) but also in parts of the developed world (e.g., the drylands of the American Southwest) (Lambin, Geist, and Lepers 2003). In the absence of serious efforts to enhance the flexibility of regimes under conditions of this sort, we can expect to encounter emergent patterns that exemplify what I have called arrested development or collapse.

When regimes are more flexible, on the other hand, different institutional dynamics come into play. The case of flexible regimes operating in stable settings may create situations in which there is excess institutional capacity that can be held in reserve for use in addressing future changes in biophysical or socioeconomic conditions. Some observers are likely to seize on situations of this kind as evidence that governance systems are epiphenomena, changing in response to minor fluctuations in the preferences of powerful actors or slight shifts in political alliances (Strange 1983; Mearsheimer 1994/1995). Yet there is a difference between adaptability and a lack of institutional teeth.

A more interesting combination, occupying the lower-right-hand cell of the matrix, encompasses situations in which flexible or adaptive regimes operate in volatile settings. The key question here is whether regimes are resilient in the sense that they have the capacity to make significant changes to cope with external stresses in a manner that leaves their fundamental character unchanged (Holling and Gunderson 2002). Among the cases I have examined in depth, the Antarctic Treaty System stands out as an arrangement that has confronted recurrent changes in the relevant socioeconomic setting and consistently found ways to cope with these stresses that have left the defining features of the regime intact. Another familiar case that belongs in this category is the regime dealing with transboundary air pollution based on the 1979 Geneva Convention on Long-range Transboundary Air Pollution (Chossudovsky 1988; Soroos 1997). The secret to success in this case has been the addition of a number of substantive protocols designed to build out the regime to address emerging problems relating to fluxes of contaminants, including volatile organic compounds, heavy metals, and persistent organic pollutants. The regime remains recognizable to those who crafted the terms of the 1979 convention. But it now encompasses a range of pollutants that is far broader than the original emphasis on sulfur dioxide and nitrogen oxides.

Both empirical observations and analytical inferences confirm the importance of the alignment between endogenous and exogenous factors as a key determinant of emergent patterns in international environmental governance. We are at an early stage in our efforts to understand these patterns. The evidence I have drawn on is largely qualitative and, at least in some cases, subject to a variety of interpretations. Still, the endogenous-exogenous alignment thesis seems robust. Regimes do change in ways that reflect the character of this alignment, whether the result is progressive development as in the case of the ozone regime, punctuated equilibrium as in the case of the Antarctic Treaty System, or growing stresses leading to collapse as in the second incarnation of the regime for northern fur seals. It seems likely that a further exploration of this relationship will yield more nuanced but broadly compatible insights.

Additional Patterns

Are there emergent patterns in environmental governance not captured in the experiences of this book's case studies? The five patterns I have examined do not constitute a proper taxonomy in the sense of a set of alternatives that are mutually exclusive and exhaustive with regard to some well-defined dimension or theme of theoretical interest. I have identified the five patterns analyzed in the previous chapters inductively on the basis of a long-standing engagement in research on international environmental regimes. These patterns cover a wide range of cases. But I have not made a sustained effort to find additional patterns that have occurred in the past or that may occur in the future whether or not we can find examples in the existing universe of international environmental regimes. As we take the next steps in the study of institutional dynamics, it seems worth exploring the prospect that other patterns of institutional change may come to the surface in various settings.

One possibility features a more or less extended period marked by little or no progress until a tipping point occurs that triggers a takeoff, a rapid transition to either progressive development or punctuated equilibrium. The idea here resembles accounts of economic development that focus on the phenomenon of takeoff leading to a period of sustained growth (Rostow 1971). Analytically, we can treat this phenomenon as a process in which a number of conditions are necessary for institutional development and there is little visible progress until all the preconditions for progressive development or punctuated equilibrium are met. A set of individually necessary and collectively sufficient conditions would control change in such cases. Central to this line of thinking is the idea of institutional takeoff. Once a tipping point is reached or a critical threshold crossed, on this account, a regime may surge ahead rapidly to achieve a level of effectiveness that represents a qualitative shift from the lackluster performance characteristic of the period preceding the tipping point. We may think of such occurrences as state changes or regime shifts in which governance systems turn a corner moving from lackluster performance to a new condition featuring enhanced performance.

Can we find real-world examples of this pattern of *takeoff to sustained development*? Unambiguous examples are hard to come by in the existing universe of international environmental regimes. Some may detect such a pattern in the development of the OSPAR regime based on the Convention for the Protection of the Marine Environment of the North-East Atlantic, particularly with respect to pollution in the North Sea, or in the regime dealing with long-range transboundary air pollution in Europe. But the evidence in both these cases seems debatable. Climate optimists hope for some such transition in the case of the climate regime operating today under the provisions of the UN Framework Convention on Climate Change and the 1997 Kyoto Protocol. As I argue in chapter 4, it is hard to avoid the conclusion that the climate regime has conformed to the pattern I call arrested development during the period from the signing of the climate convention at the 1992 UN Conference on Environment and Development to the present. This is especially true once we take into account the growing consensus in the scientific community that we have no more than one or two decades to initiate decisive steps needed to avoid drastic changes in the Earth's climate system (Linden 2006). Are the hopes of the optimists realistic in this setting? It is never easy to evaluate how close we are to important thresholds or tipping points with regard to specific governance systems. It is hard to make a compelling case for the proposition that we are about to reach such a threshold in climate governance. Current efforts to reach agreement on the terms of a replacement for the Kyoto Protocol to take effect in 2013 do not inspire confidence in these terms. Still, we often fail to foresee the approach of such thresholds, an observation that suggests we should devote some systematic thought to identifying appropriate responses should a window of opportunity open in the near future to make fundamental changes in the character of the climate regime.

A second candidate for an additional pattern of change features the termination of institutional arrangements that are so successful that they work their way out of a job and, as a result, go dormant or disappear entirely. Perhaps the most striking case in recent times is the trusteeship system created under the terms of the UN Charter as a successor to the mandate system developed during the 1920s under the auspices of the League of Nations. The trusteeship system has completed its assigned task; virtually all the trust territories identified in the aftermath of World War II have become independent nation states or moved out of the category of trust territories in some other fashion (Wilde 2007). The Trusteeship Council remains in existence as one of the principal organs of the United Nations. But its job is essentially done. This is the backdrop for proposals voiced in several quarters calling for a transformation of the Trusteeship Council into a body with a clear mission to protect the Earth system from the destructive impacts of human actions.

Perhaps more relevant to the issues discussed in this book are regimes dealing with human health that succeed so well that they are no longer needed (Cooper 1989; Cooper, Kirton, and Schrecker 2007). An excellent example is the campaign waged in the period following World War II to eradicate smallpox on a global scale. With the fulfillment of this goal in the 1970s, the apparatus assembled to implement or oversee efforts to wipe out smallpox became obsolete. Can we imagine similar cases of what might be called a pattern of *successful termination* in dealing with international environmental problems? The ozone regime could become a candidate for this pattern. Once all ozone-depleting substances are banned, there are no remaining producers of the relevant chemicals, and the problem of ODS banks is resolved, the need for a regime to protect stratospheric ozone from such substances may evaporate. We are not there yet in the case of stratospheric ozone. For the most part, the need for regimes dealing with large-scale environmental issues (e.g., climate change, the loss of biodiversity, or desertification) will be ongoing.

A third candidate involves situations in which two or more regimes that have arisen independently and that are rooted in separate agreements are subsequently joined together in the interests of forming a more effective governance system to address a multidimensional problem. Regimes generally form in response to specific and well-defined problems. I have examined regimes that address problems involving fur seals, whales, stratospheric ozone, and so forth. In the real world, however, problems are seldom neatly packaged into issues that do not interact with one another. Climate change is a major factor affecting efforts to protect biological diversity. The regime for international trade has important consequences for efforts to build regimes designed to protect various species and stocks of whales. Sometimes, the best that can be done is to find ways to alleviate or manage the resultant interplay between or among separate regimes. But there are cases in which the relevant parties reevaluate the roles of distinct regimes and take steps to consolidate these arrangements in the interests of improving their performance in solving problems.

A clear example of this pattern of *consolidation* is the integration of the 1972 Oslo Convention aimed at the dumping of pollutants at sea and the 1974 Paris Convention on the prevention of pollution from land-based sources to produce the 1992 OSPAR Convention. A somewhat more ambiguous case involves the development of closer links among the various international agreements that deal directly or indirectly with plant genetic resources. Unlike the case of OSPAR, a number of separate arrangements that are relevant to the protection of plant genetic resources remain in place. But those who have looked at the dynamics of this interrelated set of arrangements have concluded that what has emerged in recent years in this realm deserves to be characterized as a regime complex (Raustiala and Victor 2004).

Yet another potential pattern of change arises when international regimes dealing with focused concerns (e.g., fisheries management) require substantial adjustment in the wake of changes in broader or overarching institutional arrangements. Perhaps the most striking recent example of this pattern of institutional restructuring centers on reforms in marine regimes needed to accommodate overarching changes codified in the 1982 UN Convention on the Law of the Sea and, specifically, the creation of Exclusive Economic Zones (EEZs) as a major feature of the new law of the sea (Ebbin, Hoel, and Sydnes 2005). This development precipitated a cascade of changes in preexisting arrangements creating a pattern of change that may be called *institutional adjustment*. Prominent examples include regional fisheries management organizations (RFMOs), such as the Northwest Atlantic Fisheries Organization and the Northeast Atlantic Fisheries Organization (Stokke 2001). In both cases, far-reaching changes occurred following the adoption of the 1982 convention. These RFMOs did not collapse and disappear, but they did undergo major changes needed to bring them in line with the new law of the sea.

Could something like this happen in other areas? An interesting candidate involves the prospect of the development of a new law of the atmosphere. As things stand now, we have a growing collection of piecemeal arrangements covering specific atmospheric concerns including ozone-depleting substances, acid rain, emissions of greenhouse gases, air traffic control, the use of the geomagnetic spectrum, celestial bodies, and even the demilitarization of space (Soroos 1997). But we have no comprehensive or overarching law of the atmosphere analogous to the law of the sea. Is there a role for such a comprehensive law of the atmosphere? Would the development of a comprehensive governance system in this realm be politically feasible, even if it became clear that it would be desirable? We have no way of arriving at explicit answers to these questions at this juncture. But it seems clear that the emergence of such a comprehensive law of the atmosphere would necessitate considerable adjustments in existing regimes focused on specific issues like transboundary air pollution, ozone depletion, and climate change.

Underlying Causes

Are there underlying causes of change that are not evident in the accounts of the cases examined in the preceding chapters but that nonetheless play important roles behind the scenes? Many observers have drawn a distinction between proximate and underlying causes in efforts to account for the creation of international environmental regimes as well as the effectiveness of such arrangements once they are in place (Krasner 1983a). Factors relating to systems of property rights and management practices figure prominently in analyses of the forces leading to land degradation or the depletion of marine fisheries. But these are proximate causes in comparison with the growth of human populations putting pressure on wildlife and fish stocks or the introduction of new technologies allowing farmers to use fertilizers and pesticides to make their land more productive and permitting industrial fishers to sweep up living organisms from the sea floor. The operation of these underlying causes need not be regarded as diminishing the importance of understanding the proximate causes, especially in cases where prospects for regulating proximate causes like systems of property rights are considerably brighter than prospects for altering underlying causes. But it is important to bear in mind the importance of underlying causes in constructing explanations of emergent patterns in environmental governance as well as in seeking to improve the performance of existing regimes.

Can we make good use of this distinction as we move from mainstream efforts to understand regime formation and effectiveness to the study of institutional dynamics? The answer to this question is in the affirmative.

One way to approach this topic is to ask about the impacts of power, interests, and ideas, three clusters of forces widely treated as underlying causes among political scientists and others who seek to understand processes of social choice or patterns of collective behavior (Hasenclever, Mayer, and Rittberger 1997).

Understanding shifts in the underlying structure of power in international society is undoubtedly important to any study of institutional dynamics regarding specific issue areas like human-environment relations. The rise of a dominant power or a hegemon may be good news when it turns the collection of actors concerned with a particular environmental problem into a privileged group with regard to the supply of institutional changes treated as public goods (Olson 1965). But the emergence of a hegemon also can be a source of bad news when the dominant actor endeavors to impose its will regarding the substance of a regime or behaves as a laggard unwilling to play an active role in efforts to come to terms with issues like climate change, desertification, or the depletion of marine fisheries. An examination of the role of the United States regarding the development of cooperative arrangements in a number of issue areas yields clear-cut examples of such behavior. When the United States is engaged (e.g., in the case of efforts to protect stratospheric ozone), progress is steady even though specific responses to collective-action problems reflect the preferences of the dominant actor. When the hegemon drags its feet or refuses to participate actively, on the other hand, the emergence of the pattern of institutional change I call arrested development is probable. The actions of the United States have become a drag on the efforts of those seeking to implement the law of the sea or to push the climate regime past a tipping point leading to progressive development. The details will vary from case to case. But the basic point is clear. All regimes operate in broader socioeconomic settings in which the underlying structure of power plays an important role in shaping outcomes in specific cases. In seeking to explain patterns of change that occur in specific cases (e.g., the contrast between progressive development in the case of ozone and arrested development in the case of climate change), it always helps to identify the geopolitical forces at work and to think about the effects of the underlying structure of power in shaping observed patterns of institutional change.

The role of interests, treated as a second cluster of underlying causal forces, is another major concern in efforts to understand institutional dynamics. Sometimes interests manifest themselves in terms that are specific to individual issue areas. The fact that there is an asymmetry between those responsible for the buildup of greenhouse gases in the atmosphere and those likely to become early victims of climate change is important to any effort to explain the occurrence of arrested development in this case. Likewise, the fact that the United States and other members of the so-called Miami Group are interested in exporting genetically modified crops is relevant to any effort to forecast the likely fate of the 2000 Biosafety Protocol to the Convention on Biological Diversity. But there are broader patterns of interests extending beyond the realm of specific issues that come into play in efforts to deal with a variety of issues. The differences between advocates of decarbonization and those content to rely on the development of technologies that can sequester carbon already in the atmosphere reflects a profound difference in interests. There are legitimate—though typically inconclusive—debates about specific issues like the performance of the Clean Development Mechanism or the role of emissions trading schemes in efforts to reduce overall emissions of greenhouse gases. But many arguments for and against specific instruments are easier to explain in terms of the broader or overarching interests of their proponents than in terms of compelling evidence regarding their likely effects under the conditions prevailing in specific cases. The broad and persistent preference of the United States for measures that minimize the regulatory role of the state in efforts to come to terms with climate change exemplifies this link between broader patterns of interests and the character of the provisions of individual arrangements like the current climate regime.

It is a short step from this discussion of the role of preferences to a consideration of the role of ideas in explaining patterns in international environmental governance. Others have demonstrated the importance of shifts in discourses or in prevailing paradigms in processes leading to the formation of environmental regimes (Litfin 1994). A particularly prominent case in point involves the role of the newly emerging discourse of Earth system science (Steffen et al. 2004) in the processes eventuating in the adoption of the Montreal Protocol on ozone-depleting substances in 1987. Similar relationships can be seen with regard to patterns of institutional change in existing regimes. The rise of the concept of ecosystem-based management is clearly visible in the processes leading to the addition of the 1980 Convention on the Conservation of Antarctic Marine Living Resources to the ATS; much the same is true of the influence of the idea of wilderness in the framing of the 1991 Environmental Protocol to the Antarctic Treaty. As the cases of the regimes for whales and whaling and the northern fur seal make clear, broader disagreements regarding how we think about human-environment relations can and often do spill over and become encapsulated in the institutional dynamics of specific regimes. In the case of whales, the regime has become a battleground between the views of those who think in terms of consumptive uses that are sustainable and those advocating a right to life for individual animals. The occurrence of the pattern I call diversion followed by gridlock in this case is largely a reflection of this battle; the fact that no resolution to this battle is currently in sight reflects the larger standoff between proponents of sustainable use and animal rights. The collapse of the fur seal regime, by contrast, is a case in which the regime failed to adjust to the shift from a perspective emphasizing maximum sustainable yields to the paradigm of ecosystem-based management emerging in the broader community of those concerned with human-environment relations. It is notable that the new convention on Antarctic marine living resources adopted in 1980 features innovative thinking relating to ecosystem-based management, whereas the preexisting fur seal regime was unable to integrate the precepts of ecosystem-based management when the 1957 convention came up for renewal in the mid-1980s. It may well be that it is harder to restructure existing regimes to capture shifts in larger perspectives on human-environment relations than to incorporate such changes into the provisions of newly created regimes. The fur seal case suggests that changes in overarching perspectives are not easy to assimilate into existing governance systems. The result, especially in cases where the regime itself is rigid or inflexible, can be a transition leading to total collapse (e.g., the fur seal regime) or to a situation in which the regime becomes the locus of a protracted and inconclusive confrontation between proponents of alternative paths to reform (e.g., the regime for whales and whaling).

Research Frontiers

What cutting-edge questions offer attractive targets of opportunity for those interested in participating in the next phase of research on institutional change? As I argue throughout this book, the analysis of institutional change is less advanced than research on regime formation and institutional effectiveness, at least in the realm of international environmental governance. For those looking for research opportunities, this is good news. Institutional change is a fundamental feature of environmental governance systems, but research on institutional dynamics is an infant industry. To explore the implications of this proposition, I group the cutting-edge issues in this area into two categories.

The first category arises from the observation that governance systems, like the biophysical and socioeconomic systems they seek to steer, are complex and dynamic systems. This opens a line of questions dealing with the themes of robustness, resilience, and state changes as they pertain to international environmental regimes. An attractive feature of this perspective is the introduction of a distinction between constitutive changes and changes that alter operating rules while leaving constitutive arrangements unchanged. I then turn to a second and larger set of developments involving global environmental changes and global social changes and introduce a line of inquiry concerning the implications of these changes in the character of the broader setting for the dynamics of international environmental regimes. The result is an account that points to a new class of forces leading to institutional change as well as to some implications for policy that I take up in the concluding section of this chapter.

Turning first to the observation that regimes are complex and dynamic systems, we can start with the distinction between robustness and resilience (Anderies, Janssen, and Ostrom 2004). Robust regimes have the capacity to handle a wide range of challenges without undergoing significant changes in their operating rules. Resilience, by contrast, centers on the capacity of governance systems to introduce changes in operating rules to cope with unforeseen or unanticipated challenges, without changing constitutive provisions or experiencing transformative change (Holling and Gunderson 2002). Robust regimes perform well without generating a need for substantive adjustments. From the perspective of efforts to understand patterns of change in environmental regimes, this constitutes the null case. An assessment of resilience, on the other hand, must start with a distinction between operating rules and constitutive elements. It seems uncontroversial to say that the regime for stratospheric ozone has experienced considerable change in operating rules, including changes in phaseout schedules for specific chemicals, the addition of the Multilateral Fund to make participation in the regime attractive to developing countries, and the extension of the regime to additional families of chemicals. But cases like the Antarctic Treaty System and the regime for whales and whaling are more difficult to characterize in these terms. Given the fact that the ATS started as an arrangement designed to solve jurisdictional disputes and to demilitarize Antarctica, it is clear that the rise of environmental concerns in this regime and especially the new provisions incorporated into the regime in the form of the 1991 Environmental Protocol have brought about a watershed in the character of this governance system. The case of the regime for whales and whaling is even harder to evaluate in these terms. Many states that voted for the moratorium in 1982 thought of this action as a temporary measure needed to rebuild whale stocks and to develop the Revised Management Procedure to regulate commercial whaling when it resumed in the future. But given the shifting composition of the regime's membership and the nature of its decision rules, what began as a time-out evolved into a de facto constitutive change favoring the preferences of the antiwhaling community and undermining the position of those seeking to renew commercial whaling under the Revised Management Procedure. In effect, the regime experienced a state change that no one had foreseen clearly.

In the language of those who think in terms of complex and dynamic systems, state changes involve nonlinear shifts normally moving the system from one basin of attraction to another (Walker and Salt 2006). Is this perspective helpful in thinking about change in international environmental regimes? There is little evidence that regimes experience the sort of adaptive cycle that analysts who focus on biophysical or socio-ecological systems often emphasize (Gunderson and Holling 2002). In some cases, there is no evidence of the occurrence of nonlinear change. As I argue in chapter 2, the ozone regime has not undergone change of this sort. This is what justifies the use of the term progressive development to describe the pattern of change occurring in this case. But patterns of change that deserve to be characterized as nonlinear are not unusual in the realm of environmental governance. The collapse of the fur seal regime in 1984 and 1985 is a dramatic example. It is fair to characterize change in the regime for whales and whaling as nonlinear, too, though what happened in this case resulted from an action presented as a temporary measure. Many who voted for the moratorium expected that a ban on whaling for ten years would strengthen the regime as a regulatory arrangement governing the harvesting of whales rather than precipitating a state change from consumptive use to preservationism as the guiding discourse for this governance system. But this does not alter the fact that the adoption of the moratorium in 1982 touched off a dynamic leading to constitutive change.

A number of environmental regimes rest on a set of premises about the relevant biophysical and socioeconomic settings that have become outmoded as we move deeper into the Anthropocene. Implicit if not explicit are assumptions to the effect that changes in the broader setting will take the form of processes that are linear, gradual, and reversible. But changes in the real world are often nonlinear, sometimes abrupt, and frequently irreversible (Young forthcoming). The Earth's climate system is arguably the extreme case in these terms. But many other systems, including marine systems in which commercial or industrial fishing is an important activity and terrestrial systems in which deforestation or soil depletion is a major source of change, are also subject to abrupt changes whose impacts turn out to be irreversible.

What are the implications of this development for the analysis of institutional change? Regimes created to deal with turbulent biophysical and socioeconomic systems will be especially susceptible to the patterns of change I have labeled diversion and collapse. To avoid this fate, those who create environmental regimes must make a concerted effort to harness reflexivity, provide for adaptive management, and find ways to cope with uncertainty on a routine basis. Because humans are capable of anticipating future occurrences and taking steps to increase or decrease the probability of their occurrence, we can design institutional arrangements in such a way as to deal with these factors. Early-warning systems and a capacity to adapt to changes in the broader setting quickly and efficiently are important features of regimes designed to operate successfully in turbulent settings. Even so, uncertainty remains a critical condition affecting regimes that must operate in such settings. Because there is no way to avoid uncertainty in such settings, regimes must develop decision rules (e.g., the precautionary principle), implicitly if not explicitly, that can provide administrators with guidance needed to make decisions under uncertainty (Tversky and Kahneman 1974). Further analysis of the relationship between turbulence in the broader setting and the constitutive features and operating rules of environmental regimes should prove rewarding as we endeavor to improve our understanding of institutional dynamics in this realm.

Policy Relevance

Does the study of institutional dynamics yield insights that are relevant to policy and useful to policy makers dealing with specific socioecological systems? We are not now and may never be in a position to provide policy makers with surefire recipes that will yield regimes that lead to progressive development or punctuated equilibrium and that steer clear of arrested development, diversion, and even collapse. But research on institutional change does yield insights that policy makers will find helpful in thinking about the performance of specific regimes. Here, I draw a distinction between insights relevant to the design stage and insights likely to prove helpful to those charged with the management and administration of regimes once they are put in place.

A simple recognition of the dynamism of environmental regimes is a step in the right direction. There is a pronounced tendency in thinking about institutional design to assume, at least implicitly, that arrangements that seem attractive in design terms will not only prove relatively easy to implement but also remain effective indefinitely. The study of implementation is now a growth industry among researchers interested in environmental governance. We are alert to the distinction between the rules on paper and the rules in use (Ostrom 1990). So far, we have paid much less attention to emergent patterns in environmental governance systems. We now know that such systems can and often do experience significant changes brought about by both endogenous processes and exogenous forces. But we have yet to apply this knowledge in a systematic way to either the design or the administration of specific regimes.

An assessment of institutional dynamics has a number of implications that designers can and should consider. A particularly important example is the issue of the stringency of rules and procedures governing intentional adjustments of institutional arrangements in order to reflect changes in broader biophysical and socioeconomic settings. A regime that changes in response to every ripple in the broader setting will become an epiphenomenon; it will have no capacity to steer or influence the trajectory of collective outcomes in the relevant issue area. Conversely, a regime that is impossible to reform even in the wake of far-reaching changes in the relevant setting will run the risk of becoming irrelevant and ending up as a dead letter; excessive stringency is likely to eventuate in the pattern I call arrested development or even in the pattern I label collapse. There is no magic formula that can guide the designer in selecting optimal procedures for amending or revising regimes operating in specific settings. But the study of institutional change can help in identifying best practices in this realm.

Designers also need to consider rules governing membership, decision-making procedures, revenue streams, and dispute settlement. All these are matters that research on institutional change can illuminate. Rules setting forth requirements for the acceptance of new members and the distinction between consultative parties and others appear to constitute a strength in the case of the ATS. The absence of such rules in the case of the regime for whales and whaling played a part in opening up opportunities for manipulation that both antiwhalers and prowhalers have sought to exploit to their own advantage (DeSombre 2001). Decision rules are clearly relevant to the dynamics of institutional change. The existence of relatively flexible decision rules has played a role in the development of the pattern of progressive development in the case of ozone depletion. The decision rule in use in the regime for whales and whaling is a source of the paralysis that has characterized this regime in recent years. Most international environmental regimes suffer from a lack of secure sources of funding. The existence of the Multilateral Fund has played a role in the pattern of progressive development occurring in the case of the ozone regime. The absence of adequate funds is a source of the arrested development that has afflicted the climate regime. The absence of courts with compulsory jurisdiction is a problem for all regimes. But more limited arrangements, like the ozone regime's noncompliance procedure, can help in adapting institutional arrangements to exogenous changes (Victor 1998). None of this yields a simple blueprint that will assure success in the design of environmental regimes. But an analysis of the links between design features and the patterns of institutional change likely to occur in specific cases can be an important source of insights for those crafting the provisions of agreements establishing environmental regimes in a variety of issue areas.

Knowledge of institutional dynamics can also prove helpful to those charged with managing or administering regimes once they are up and running. The application of such insights on a day-to-day basis is easier said than done. Given the shifts in the broader socioeconomic setting, the biophysical condition of the Bering Sea ecosystem, and prevailing discourses, there is probably little that administrators of the fur seal regime could have done to maintain the viability of this arrangement during the 1980s. On the other hand, the achievement of progressive development was relatively easy in the case of the ozone regime, given the flexibility of the regime itself and the ease with which major producers were able to come up with alternatives to products banned as a result of the phaseout provisions of the Montreal Protocol as accelerated by decisions reached from time to time at the annual meetings of the Conference of the Parties and the Meeting of the Parties.

Still, administrators will take a lively interest in the growth of knowledge about institutional dynamics, especially as they seek to devise operating rules needed to carry out a regime's provisions on a day-to-day basis. Research on policy instruments offers a range of insights in these terms. What works in one case (e.g., the use of phaseout schedules to move the system toward a full prohibition on the production and consumption of ozone-depleting substances) may not work in other cases (e.g., the climate regime where the challenge is to regulate emissions of greenhouse gases rather than to prohibit them altogether). There is much to be learned about the use of incentive mechanisms, such as emissions trading schemes and the Clean Development Mechanism in the case of climate, and the development of techniques for valuing ecosystem services, such as contributions to the maintenance of biological diversity, that are not captured in market prices. What is at stake here is the capacity of administrators to make adjustments in operating rules in the interests of enhancing the resilience of specific regimes, even when they are not—and should not be—in a position to alter the constitutive provisions of the governance systems they serve. The trick is to make midcourse corrections or to engage in fine-tuning in such a way as to maximize the prospects of entraining the patterns of change I call progressive development and punctuated equilibrium and to avoid falling into the traps of arrested development or diversion, much less triggering chains of events that take us in the direction of collapse. Administrators frequently operate in circumstances that offer little flexibility to engage in the sort of institutional adjustment under consideration here; they are often so risk averse that they are unwilling to take a chance on adjustments that may not work and that, in any case, may be regarded by senior policy makers as exceeding the bounds of their authority. Still, recognition of the fact that environmental regimes are highly dynamic systems makes it important for administrators to learn how to avoid disruptive regime shifts and how to take advantage of windows of opportunity during which it is feasible to introduce new practices that can enhance the performance of governance systems.

Conclusion

I have endeavored to identify and explain emergent patterns in international environmental governance. Some of these patterns (progressive development and punctuated equilibrium) are more promising than others (arrested development, diversion, and collapse) from the point of view of those seeking to solve problems arising in human-environment relations, such as the depletion of renewable resources, the seasonal thinning of stratospheric ozone, or changes in the Earth's climate system. There is variation as well in the extent to which it is possible to pinpoint the drivers of institutional change and, as a result, to fashion governance systems that are likely to prove successful in terms of problem solving. In cases like climate change where the character of the problem makes it unusually difficult to find a solution, understanding institutional change may not help much in the search for problem-solving strategies that are feasible as well as desirable. But even in such cases, getting decision rules and policy instruments right may go some distance toward preparing the ground for what I call progressive development in contrast to arrested development or collapse.

This study is based on an analysis of emergent patterns in international environmental governance. Strictly speaking, there is no basis for generalizing the findings I have reported in this final chapter beyond the confines of that universe of cases. But it is apparent that all institutions are dynamic systems and that the issue of emergent patterns is relevant to governance systems addressing a wide range of issues and operating at all levels of social organization. I make no formal claims regarding the applicability of my findings (e.g., the robustness of the endogenous-exogenous alignment thesis) to the study of institutional dynamics in other domains. But it strikes me that there is ample scope for a constructive dialog among those seeking to understand both emergent patterns in governance systems and the mechanisms that give rise to these patterns across a range of issue areas and levels of social organization.

Notes

Chapter 1

1. The distinction between robustness and resilience is a significant and relatively new addition to the literature on complex and dynamic systems (Anderies, Janssen, and Ostrom 2004), but it does not alter the defining characteristics of this mode of analysis.

2. The terms *disturbance* and *threat* are used in this literature in a manner that is largely synonymous with the term *stress*. To avoid confusion, I usually speak of stress in the analysis to follow.

Chapter 3

1. Many commentators regard this as reflecting a lack of agreement rather than a lack of interest in environmental and resource issues. But the record is not conclusive on this point.

2. The twelve original signatories assume they will remain consultative parties, even if their Antarctic research programs wane.

3. Information on the IPY is available at www//ipy.org.

4. Under the terms of the 1982 UN Convention on the Law of the Sea, parties must file claims to areas of the continental shelf lying beyond their Exclusive Economic Zones within a period of ten years following ratification. A number of countries have included such claims to areas surrounding Antarctica as a part of their general claims to jurisdiction over extended areas of the continental shelf. But for the most part, the inclusion of areas adjacent to Antarctica in these claims has been pro forma or motivated by a desire to cover future contingencies.

Chapter 4

1. Data on GHG emissions through 2006 are available at the UNFCCC Secretariat Web site (http://unfccc.int/).

2. Between 2003 and 2004, EU-15 emissions actually increased by 0.3 percent.

3. The fifteenth conference of the parties (COP 15) of the UNFCCC took place in Copenhagen, Denmark, December 7–19, 2009. The good news about this event is that it raised the profile of climate change on the international policy agenda. The presence of 115 heads of state and heads of government plus the UN Secretary General at the high-level segment of the conference provides clear evidence that many now take this issue seriously and treat it as a matter requiring attention on an urgent basis. This offers some basis for hope for future efforts to tackle the problem of climate change and, in the process, to meet the overall goal of the climate regime. Yet COP 15 failed to produce the step-level change needed to break the pattern of arrested development that has plagued this regime for many years. The principal product of the conference, known as the Copenhagen Accord, is a political agreement negotiated by the leaders of the United States, China, India, Brazil, and South Africa and "noted" by conference participants at their final plenary session December 18–19. This agreement acknowledges the need to hold temperature increases below 2°C to avoid dangerous changes in the climate system, calls on both Annex 1 and non-Annex 1 countries to make explicit pledges regarding emissions limitations, contains promises relating to financial support for mitigation and adaptation, addresses the need for reporting mechanisms, and recognizes the role of reducing emissions from deforestation and forest degradation. But the Copenhagen Accord is not a legally binding agreement, and it does not specify any target date for reaching agreement on a legally binding successor to the Kyoto Protocol or other supplement to the UNFCCC itself. Moreover, the accord does not spell out overall targets and timetables for future emissions reductions, and it offers few details regarding the nature of the financial arrangements or reporting mechanisms to be established to carry forward the work of the regime. It is important to acknowledge and respect the hard work of all those who participated in COP 15. But the results provide no basis for revising the analysis set forth in this chapter.

Chapter 6

1. William Seward, the U.S. secretary of state, negotiated the agreement on behalf of the United States. Many critics of the day regarded this as a bad bargain and ridiculed Seward's decision to purchase Alaska on behalf of the United States.

2. What had been a population of millions of animals dwindled to 150,000 or fewer by the signing of the 1911 convention.

3. The regime makes an exception for indigenous peoples who are allowed to take seals at sea, provided they do so in a traditional manner (Art. IV of the 1911 convention; Art. VII of the 1957 Interim Convention). This practice has never reached a level where it would affect the status of the fur seal herds.

References

Adger, W. Neil. 2006. Vulnerability. *Global Environmental Change* 16:268–281.

Aldy, Joseph E., and Robert N. Stavins, eds. 2007. *Architectures for Agreement: Addressing Global Climate Change in the Post-Kyoto World*. New York: Cambridge University Press.

Alley, Richard. 2000. *The Two-Mile Time Machine: Ice Cores, Abrupt Climate Change, and Our Future*. Princeton: Princeton University Press.

Allison, Graham. 1971. *Essence of Decision*. Boston: Little Brown.

Anderies, John M., Marco A. Janssen, and Elinor Ostrom. 2004. A Framework to Analyze the Robustness of Social-ecological Systems from an Institutional Perspective. *Ecology and Society* 19 (1) article 18.

Andersen, Stephen O., and K. Madhava Sarma. 2002. *Protecting the Ozone Layer: The United Nations History*. London: Earthscan.

Andresen, Steinar. 1989. Science and Politics in the International Management of Whales. Marine Policy 13:99–117.

Andresen, Steinar. 1998. The Making and Implementation of Whaling Policies: Does Participation Make a Difference. In *The Implementation and Effectiveness of International Environmental Commitments,* edited by David G. Victor, Kal Raustiala, and Eugene B. Skolnikoff. Cambridge: MIT Press, 431–474.

Andresen, Steinar. 2002. The Whaling Regime: The International Convention for the Regulation of Whaling (ICRW) and the International Whaling Commission. In *Science and International Environmental Regimes: Integrity and Involvement*, edited by Steinar Andresen, Tora Skodvin, Arild Underdal, and Jorgen Wettestad. Manchester: University of Manchester Press, 61–86.

Arctic Council. 2004. *Arctic Climate Impact Assessment.* Cambridge: Cambridge University Press.

Aron, William. 2001. Science and the IWC. In *Toward a Sustainable Whaling Regime*, edited by Robert L. Friedheim. Seattle: University of Washington Press, 105–122.

Baden, John A., and Douglas Noonan, eds. 1998. *Managing the Commons*, 2nd ed. Bloomington: Indiana University Press.

Baumgartner, Frank R., and Bryan D. Jones. 1993. *Agendas and Instability in American Politics*. Chicago: University of Chicago Press.

Beck, Peter J. 1986. *The International Politics of Antarctica*. New York: St. Martin's Press.

Belanger, Dian Olson. 2006. *Deep Freeze: The International Geophysical Year, and the Origins of Antarctica's Age of Science*. Boulder: University Press of Colorado.

Benedick, Richard. 1991/1998. *Ozone Diplomacy: New Directions in Safeguarding the Planet*, 2nd ed. Cambridge: Harvard University Press.

Berkes, Fikret, and Carl Folke, eds. 1998. *Linking Social and Ecological Systems: Management Practices and Social Mechanisms for Building Resilience*. Cambridge: Cambridge University Press.

Berkman, Paul. 2002. *Science into Policy: Global Lessons from Antarctica*. London: Academic Press.

Betsill, Michelle M., and Barry G. Rabe. 2009. Climate Change and Multi-Level Governance: The Emerging State and Local Roles. In *Towards Sustainable Communities*, 2nd ed., edited by D. A. Mazmanian and M. E. Kraft. Cambridge: MIT Press, 201–226.

Bolin, Bert. 1997. Scientific Assessment of Climate Change. In *International Politics of Climate Change: Key Issues and Critical Actors*, edited by Gunnar Fermann. Oslo: Scandinavian University Press, 83–109.

Breitmeier, Helmut, Oran R. Young, and Michael Zürn. 2006. *The Analysis of International Environmental Regimes: From Case Study to Database*. Cambridge: MIT Press.

Bryner, Gary. 1993. *Blue Skies, Green Politics: The Clean Air Act of 1990*. Washington, D.C.: CQ Press.

Burke, William T. 2001. A New Whaling Agreement and International Law. In *Toward a Sustainable Whaling Regime*, edited by Robert L. Friedheim. Seattle: University of Washington Press, 105–122.

Carson, Rachel. 1962. *Silent Spring*. Boston: Houghton Mifflin.

Caulfield, Richard A. 1997. *Greenlanders, Whales, and Whaling: Sustainability and Self-Determination in the Arctic*. Hanover, N.H.: University Press of New England.

Chayes, Abram, and Antonia Handler Chayes. 1995. *The New Sovereignty: Compliance with International Regulatory Agreements*. Cambridge: Harvard University Press.

Chossudovsky, Evgeny M. 1988. *"East-West" Diplomacy for Environment in the United Nations*. New York: UNITAR.

Congressional Research Service. 2008. "CRS Report for Congress- China's Greenhouse Gas Emissions and Mitigation Policies." Order Code RL34659, 10 September.

Cooper, Andrew F., John J. Kirton, and Ted Schrecker, eds. 2007. *Governing Global Health: Challenges, Responses, Innovation*. Aldershot, UK: Ashgate.

Cooper, Richard N. 1989. International Cooperation in Public Health as a Prologue to Macroeconomic Cooperation. In *Can Nations Agree? Issues in International Economic Cooperation*, edited by Richard N. Cooper, Barry Eichengreen, C. Randall Henning, Gerald Hothman, and Robert D. Putnam. Washington, D.C.: Brookings Institution, 178–254.

Crowder, Larry B., Gail Osherenko, Oran R. Young, Satie Airamé, Elliott A. Norse, Nancy Baron, Jon C. Day, Fanny Douvere, Charles N. Ehler, Benjamin S. Halpern, Stephen J. Langdon, Karen L. McLeod, John C. Ogden, Robbin E. Peach, Andrew A. Rosenberg, and James A. Wilson. 2006. Resolving Mismatches in U.S. Ocean Governance. *Science* 313 (4 August), 617–618.

Crutzen, Paul. 2002. Geology of Mankind—The Anthropocene. *Nature* 415:23.

Day, Jon. 2002. Zoning—Lessons from the Great Barrier Reef Marine Park. *Ocean and Coastal Management* 45:139–156.

DeSombre, Elizabeth. 2000. *Domestic Sources of International Environmental Policy: Industry, Environment, and U.S. Power*. Cambridge: MIT Press.

DeSombre, Elizabeth. 2001. Distorting Global Governance: Membership, Voting, and the IWC. In *Toward a Sustainable Whaling Regime*, edited by Robert L. Friedheim. Seattle: University of Washington Press, 183–199.

Dessler, Andrew E., and Edward A. Parson. 2006. *The Science and Politics of Global Climate Change*. Cambridge: Cambridge University Press.

Diamond, Jared. 2005. *Collapse: How Societies Choose to Fail or Succeed*. New York: Viking.

Dobbs, David. 2000. *The Great Gulf: Fishermen, Scientists, and the Struggle to Restore the World's Greatest Fishery*. Washington, D.C.: Island Press.

Dolin, Eric Jay. 2007. *Leviathan: The History of Whaling in America*. New York: W.W. Norton.

Downs, Anthony. 1972. Up and Down with Ecology: The "Issue-Attention" Cycle. *Public Interest* 28:38–50.

Dryzek, John S. 1997. *The Politics of the Earth: Environmental Discourses*. Oxford: Oxford University Press.

Dryzek, John S. 2006. *Deliberative Global Politics*. Oxford: Polity Press.

Ebbin, Syma, Alf Hakon Hoel, and Are Sydnes, eds. 2005. *A Sea Change: The Exclusive Economic Zone and Governance Institutions for Living Marine Resources*. Dordrecht: Springer Verlag.

Eggertsson, Thrainn. 2005. *Imperfect Institutions: Possibilities and Limits of Reform*. Ann Arbor: University of Michigan Press.

European Environment Agency. 2008. Greenhouse gas emission trends (CSI 010): Assessment published February 2008. Available at http://themes.eea.eruopa.eu/.

Farrell, Alexander, and Jill Jäger, eds. 2006. *Assessments of Regional and Global Environmental Risks: Designing Processes for the Effective Use of Science in Decisionmaking*. Washington, D.C.: Resources for the Future.

Flannery, Timothy. 2005. *The Weather Makers: How Man Is Changing the Climate and What It Means for Life on Earth*. New York: Atlantic Monthly Press.

Fluharty, David. 2005. Evolving ecosystems approaches to management of fisheries in the USA. *Marine Ecology Progress Series* 300:248–253.

Freeman, Milton M. R. and Urs P. Kreuter, eds. 1994. *Elephants and Whales: Resources for Whom?* Basel: Gordon and Breach.

Friedheim, Robert L. 1996. Moderation in Pursuit of Justice: Explaining Japan's Failure in the International Whaling Negotiations. *Ocean Development and International Law* 27:349–378.

Friedheim, Robert L., ed. 2001a. *Toward a Sustainable Whaling Regime*. Seattle: University of Washington Press.

Friedheim, Robert L. 2001b. Fixing the Whaling Regime: A Proposal. In *Toward a Sustainable Whaling Regime*, edited by Robert L. Friedheim. Seattle: University of Washington Press, 311–335.

Friedheim, Robert L. 2001c. Negotiating in the IWC Environment. In *Toward a Sustainable Whaling Regime*, edited by Robert L. Friedheim. Seattle: University of Washington Press, 200–234.

Galaz, Victor, Per Olsson, Thomas Hahn, Carl Folke, and Uno Svedin. 2008. The Problem of Fit between Ecosystems and Governance Systems: Insights and Emerging Challenges. In *Institutions and Environmental Change*, edited by Oran R. Young, Heike Schroeder, and Leslie A. King. Cambridge: MIT Press, 147–186.

G8 Summit 2007. "Summit Declaration on Growth and Responsibility in the World Economy." Available at http://www.g-8.de/Webs/G8/EN/Homepage/home.html.

G8 Summit 2009. "Responsible Leadership for a Sustainable Future." Available at http://www.g8italia2009.it/g8/Home/.

Gay, James Thomas. 1987. *American Fur Seal Diplomacy: The Alaskan Fur Seal Controversy*. New York: Peter Lang.

Greenberg, Daniel S. 1999. *The Politics of Pure Science. New edition.* Chicago: University of Chicago Press.

Gunderson, Lance H., and C. S. Holling, eds. 2002. *Panarchy: Understanding Transformation in Human and Natural Systems*. Washington, D.C.: Island Press.

Haas, Peter M. 1992. Introduction: Epistemic Communities and International Policy Coordination. *International Organization* 46:1–35.

Haas, Peter M., Robert O. Keohane, and Marc A. Levy, eds. 1993. *Institutions for the Earth: Sources of Effective International Environmental Protection*. Cambridge: MIT Press.

Haggard, Stephan, and Beth A. Simmons. 1987. Theories of International Regimes. *International Organization* 41:491–517.

Hardin, Garrett. 1968. The Tragedy of the Commons. *Science* 162:1243–1248.

Harris, Michael. 1998. *Lament for an Ocean: The Collapse of the Atlantic Cod Fishery*. Toronto: McClelland and Stewart.

Hart, H. L. A. 1961. *The Concept of Law*. Oxford: Clarendon Press.

Hasenclever, Andreas, Peter Mayer, and Volker Rittberger. 1997. *Theories of International Regimes*. Cambridge: Cambridge University Press.

Hayek, Friedrich. 1973. *Rules and Order. Law, Legislation, and Liberty,* vol. 1. Chicago: University of Chicago Press.

Hays, Samuel P. 1975. *Conservation and the Gospel of Efficiency: The Progressive Conservation Movement 1890–1920*. New York: Atheneum.

Hays, Samuel P. 1987. *Beauty, Health, and Permanence: Environmental Politics in the United States, 1955–1985*. Cambridge: Cambridge University Press.

Hoel, Alf Hakon. 1993. Regionalization of International Whale Management: The Case of the North Atlantic Marine Mammals Commission. *Arctic* 46: 116–123.

Hoel, Alf Hakon, with contributions from Elena Andreeva, Russell Reichelt, Virginia Walsh, and Oran R. Young. 2000. "Performance of Exclusive Economic Zones." IDGEC Scoping Report No. 2.

Holling, C. S., and H. Lance Gunderson. 2002. Resilience and Adaptive Cycles. In *Panarchy: Understanding Transformation in Human and Natural Systems*, edited by Lance H. Gunderson and C. S. Holling. Washington, D.C.: Island Press, 25–62.

Homer-Dixon, Thomas. 2006. *The Up Side of Down: Creativity and the Renewal of Civilization*. Washington, D.C.: Island Press.

Intergovernmental Panel on Climate Change (IPCC) 2007. *Climate Change 2007 Synthesis Report*. Available on the IPCC Web site: http://www.ipcc.ch/.

Joyner, Christopher. 1998. *Governing the Frozen Commons: The Antarctic Regime and Environmental Protection*. Columbia: University of South Carolina Press.

Kahneman, Daniel. 2003. Maps of Bounded Rationality: Psychology for Behavioral Economics. *American Economic Review* 93:1449–1475.

Kaniaru, Donald. 2007. *The Montreal Protocol: Celebrating 20 Years of Environmental Progress*. London: Cameron May.

Kaniaru, Donald, Rajendra Shende, Scott Stone, and Durwood Zaelke eds. 2007. "Strengthening the Montreal Protocol: Insurance against Abrupt Climate Change." INECE Working Paper.

Kasperson, Jeanne X., Roger E. Kasperson, and B. J. Turner, II, eds. 1995. *Regions at Risk: Comparisons of Threatened Environments*. Tokyo: UNU Press.

Keohane, Robert O., and Joseph S. Nye, Jr. 1977. *Power and Interdependence: World Politics in Transition*. Boston: Little Brown.

Keohane, Robert O. 1984. *After Hegemony: Cooperation and Discord in the World Political Economy*. Princeton: Princeton University Press.

Keohane, Robert O., and Marc A. Levy, eds. 1996. *Institutions for Environmental Aid: Pitfalls and Promise*. Cambridge: MIT Press.

Kingdon, John W. 1995. *Agendas, Alternatives and Public Policies*, 2nd ed. New York: HarperCollins Publishers.

Klyza, Christopher McGrory, and David Sousa. 2008. *American Environmental Policy, 1990–2006*. Cambridge: MIT Press.

Krasner, Stephen D. 1983a. Structural Causes and Regime Consequences: Regimes as Intervening Variables. In *International Regimes*, edited by Stephen D. Krasner. Ithaca: Cornell University Press, 1–21.

Krasner, Stephen D., ed. 1983b. *International Regimes*. Ithaca: Cornell University Press.

Kruger, Joseph A., and William A. Pizer. 2004. Greenhouse Gas Trading in Europe: The New Grand Policy Experiment. *Environment* 46 (8):8–23.

Lambin, Eric R., Helmut J. Geist, and Erika Lepers. 2003. Dynamics of Land-Use and Land-Cover Change in Tropical Regions. *Annual Review of Environment and Resources* 28:205–241.

Lambin, Eric R., and Helmut J. Geist, eds. 2006. *Land-Use and Land-Cover Change: Local Processes and Global Impacts*. Berlin: Springer.

Larkin, Peter A. 1977. An Epitaph for the Concept of Maximum Sustainable Yield. *Transactions of the American Fisheries Society* 106:1–11.

Lee, Kai. 1993. *Compass and Gyroscope*. Washington, D.C.: Island Press.

Leopold, Aldo. 1970. The Land Ethic. In Aldo Leopold, *A Sand County Almanac with Essays on Conservation from Round River*. New York: Ballantine Books, 237–264.

Levy, Marc A., Oran R. Young, and Michael Zürn. 1995. The Study of International Regimes. *European Journal of International Relations*, 1.

Linden, Eugene. 2006. *The Winds of Change: Climate, Weather, and the Destruction of Civilizations*. New York: Simon and Schuster.

Litfin, Karen T. 1994. *Ozone Discourses: Science and Politics in Global Environmental Cooperation*. New York: Columbia University Press.

Lyster, Simon. 1985. *International Wildlife Law: An Analysis of International Treaties Concerned with the Conservation of Wildlife*. Cambridge: Cambridge University Press.

Mayewski, Paul, and Frank White. 2002. *The Ice Chronicles: The Quest to Understand Global Climate Change*. Hanover, N.H.: University Press of New England.

Meadows, Dana. 2008. *Thinking in Systems: A Primer*. White River Junction, Vt.: Chelsea Green.

Mearsheimer, John J. 1994/1995. The False Promise of International Institutions. *International Security* 19:5–49.

Miles, Edward L., Arild Underdal, Steinar Andresen, Jorgen Wettestad, and Jon Birger Skjaerseth 2002. *Environmental Regime Effectiveness: Confronting Theory with Evidence*. Cambridge: MIT Press.

Mills, Nicolaus. 2008. *Winning the Peace: The Marshall Plan and America's Coming of Age as a Superpower*. Hoboken, NJ: John Wiley and Sons.

Mirovitskaya, Natalia, Margaret Clark, and Ronald G. Purver. 1993. North Pacific Fur Seals: Regime Formation as a Means of Resolving Conflict. In *Polar Pol-*

itics: Creating International Environmental Regimes, edited by Oran R. Young and Gail Osherenko. Ithaca: Cornell University Press, 22–55.

Mitchell, Ronald, William C. Clark, David W. Cash, and Nancy M. Dickson, eds. 2006. *Global Environmental Assessments: Information and Influence*. Cambridge: MIT Press.

Morell, Virginia. 2009. Mystery of Missing Humpbacks Solved. Science 324 (29 May), 1132.

National Research Council (NRC) 1986. *Antarctic Treaty System: An Assessment*. Washington, D.C.: National Academy Press.

National Research Council (NRC). 1996. *The Bering Sea Ecosystem*. Washington, D.C.: National Academy Press.

National Research Council (NRC). 2003. *Decline of the Steller Sea Lion in Alaskan Waters: Untangling Food Webs and Fishing Nets*. Washington, D.C.: National Academy Press.

Norse, Elliott A., and Larry B. Crowder, eds. 2005. *Marine Conservation Biology: The Science of Maintaining the Sea's Biodiversity*. Washington, D.C.: Island Press.

North, Douglass C. 1990. *Institutions, Institutional Change and Economic Performance*. Cambridge: Cambridge University Press.

Oberthür, Sebastian, and Thomas Gehring, eds. 2006. *Institutional Interaction—How to Prevent Conflicts and Enhance Synergies between International and European Environmental Institutions*. Cambridge: MIT Press.

Oberthür, Sebastian, and Hermann E. Ott. 1999. *The Kyoto Protocol: International Climate Policy for the 21st Century*. Berlin: Springer.

Olson, Mancur, Jr. 1965. *The Logic of Collective Action*. Cambridge: Harvard University Press.

Olson, Mancur, Jr. 1982. *The Rise and Decline of Nations*. New Haven: Yale University Press.

Ostrom, Elinor. 1990. *Governing the Commons: The Evolution of Institutions for Collective Action*. Cambridge: Cambridge University Press.

Parson, Edward A. 2003. *Protecting the Ozone Layer: Science and Strategy*. Oxford: Oxford University Press.

Pauly, D., V. Christensen, J. Dalsgaard, R. Froese, and F. Torres, Jr. 1998. Fishing down marine food webs. *Science* 279:860–863.

Pearse, Fred. 2007. *With Speed and Violence: Why Scientists Fear Tipping Points in Climate Change*. Boston: Beacon Press.

Peterson, M. J. 1988. *Managing the Frozen South: The Creation and Evolution of the Antarctic Treaty System*. Berkeley: University of California Press.

Peterson, M. J. 1992. Whalers, Cetologists, Environmentalists, and the International Management of Whaling. International Organization 46:149–187.

Peterson, M. J. 1993. International Fisheries Management. In *Institutions for the Earth: Sources of Effective International Environmental Protection*, edited by

Peter M. Haas, Robert O. Keohane, and Marc A. Levy. Cambridge: MIT Press, 249–305.

Pikitch, E. K., Christine Santora, Elizabeth A. Babcock, Andrew Bakun, Ramon Bonfil, David O. Conover, Paul Dayton, Phaedra Doukakis, David Fluharty, Burr Heneman, Ed D. Houde, J. Link, Pat A. Livingston, Marc Mangel, Murdoch K. McAllister, John Pope, and Keith J. Sainsbury. 2004. Ecosystem-Based Fishery Management. *Science,* 305 (16 July): 346–347.

Raustiala, Kal, and David G. Victor. 2004. The Regime Complex for Plant Genetic Resources. *International Organization* 58:277–309.

Rayfuse, Rosemay G. 2007. Melting Moments: The Future of Polar Oceans Governance in a Warming World. *Review of European Community & International Environmental Law* 16:196–216.

Raymond, Leigh. 2003. *Private Rights in Public Resources: Equity and Property Allocation in Market-Based Environmental Policy*. Washington, D.C.: RFF Press.

Repetto, Robert C., ed. 2006. *Punctuated Equilibrium and the Dynamics of U.S. Environmental Policy*. New Haven: Yale University Press.

Rittberger, Volker, ed. 1990. *International Regimes in East-West Politics*. London: Pinter.

Rittberger, Volker, and Peter Mayer, eds. 1993. *Regime Theory and International Relations*. Oxford: Clarendon Press.

Rostow, Walt W. 1971. *The Stages of Economic Growth: A Non-Communist Manifesto*. Cambridge: Cambridge University Press.

Schelling, Thomas C. 1978. *Micromotives and Macrobehavior*. New York: W.W. Norton.

Sherman, Kenneth. 1992. Large Marine Ecosystems. In *Encyclopedia of Earth System Science*. vol. 2. New York: Academic Press, 653–673.

Singer, Peter. 1975. *Animal Liberation: A New Ethics for Our Treatment of Animals*. New York: Avon.

Small, George L. 1971. *The Blue Whale*. New York: Columbia University Press.

Social Learning Group. 2001. *Learning to Manage Global Environmental Risks: A Comparative History of Social Responses to Climate Change, Ozone Depletion, and Acid Precipitation*. Cambridge: MIT Press.

Soroos, Marvin. 1997. *The Endangered Atmosphere: Preserving a Global Commons*. Columbia: University of South Carolina Press.

Steffen, Will, A. Sanderson, P. Tyson, J. Jäger, P. Matson, B. Moore III, F. Oldfield, K. Richardson, H. J. Schellnhuber, B. L. Turner II, and R. J. Wasson. 2004. *Global Change and the Earth System: A Planet under Pressure*. Berlin: Springer Verlag.

Steffen, Will, Paul J. Crutzen, and John R. McNeill. 2007. The Anthropocene: Are Humans Now Overwhelming the Great Forces of Nature? *Ambio* 36 (8):614–621.

Stern, Nicholas. 2007. *The Economics of Climate Change: The Stern Review*. Cambridge: Cambridge University Press.

Stern, Paul, Oran R. Young, and Daniel Drukman, eds. 1992. *Global Environmental Change: Understanding the Human Dimensions*. Washington, D.C.: National Academy Press.

Stoett, Peter J. 1997. *The International Politics of Whaling*. Vancouver: University of British Columbia Press.

Stokke, Olav Schram, ed. 2001. *Governing High Seas Fisheries: The Interplay of Global and Regional Regimes*. Oxford: Oxford University Press.

Strange, Susan. 1983. *Cave! hic dragones:* A Critique of Regime Analysis. In *International Regimes*, edited by Stephen D. Krasner. Ithaca: Cornell University Press, 337–354.

Tainter, Joseph. 1988. *The Collapse of Complex Societies*. Cambridge: Cambridge University Press.

Tetlock, Philip E., and Aaron Belkin, eds. 1996. *Counterfactual Thought Experiments in World Politics: Logical, Methodological, and Psychological Perspectives*. Princeton: Princeton University Press.

Tietenberg, Thomas. 2002. The Tradable Permits Approach to Protecting the Commons: What Have We Learned? In *The Drama of the Commons*, edited by Elinor Ostrom, Thomas Dietz, Nives Dolsak, Paul C. Stern, Susan Stonich, Elke U. Weber. Washington, D.C.: National Academy Press, 197–232.

Toll, Richard S.J., and Gary W. Yohe. 2006. A Review of the *Stern Review. World Economy* 7:233–250.

Torrey, Barbara Boyle. 1978. *Slaves of the Harvest: The Story of the Pribilof Aleuts*. St. Paul, Alaska: Tanadgusix Corporation.

Tversky, Amos, and Daniel Kahneman. 1974. Judgment under Uncertainty: Heuristics and Biases. *Science* 185 (4157):1124–1131.

Underdal, Arild, and Oran R. Young, eds. 2004. *Regime Consequences: Methodological Challenges and Research Strategies*. Dordrecht: Kluwer Academic Publishers.

United Nations Development Programme (UNDP). 2007. *Human Development Report 2007/2008*. New York: Palgrave Macmillan.

UNFCCC Secretariat. 2007. Graphs and Figures. Accessed August 7, 2007: http://unfccc.int/files/inc/graphics/image/gif.

Velders, Guus. J. M., Stephen O. Andersen, John S. Daniel, David W. Fahey, and Mack McFarland. 2007. "The Importance of the Montreal Protocol in Protecting Climate." *Proceedings of the National Academy of Sciences,* 104 (20 March): 4814–4819.

Victor, David G. 1998. The Operation and Effectiveness of the Montreal Protocol's Non-Compliance Procedure. In *The Implementation and Effectiveness of International Environmental Commitments*, edited by David G. Victor, Kal Raustiala, and Eugene B. Skolnikoff. Cambridge: MIT Press, 137–176.

Victor, David G. 2001. *The Collapse of the Kyoto Protocol*. Princeton: Princeton University Press.

Victor, David G., Joshua C. House, and Sarah Joy. 2005. A Madisonian Approach to Climate Change. *Science* 309:1820–1821.

Vitousek, Peter, Harold Mooney, Jane Lubchenko, and Jerry Melillo. 1997. Human Domination of the Earth's Ecosystems. *Science* 277:494–499.

Walker, Bryan, and David Salt. 2006. *Resilience Thinking: Sustaining Ecosystems and People in a Changing World*. Washington, D.C.: Island Press.

Wapner, Paul. 1997. Governance in Global Civil Society. In *Global Governance*, edited by Oran R. Young. Cambridge: MIT Press, 65–84.

Warner, William. 1983. *Distant Water: The Fate of the North Atlantic Fisherman*. Boston: Little, Brown.

Wenzel, George. 1991. *Animal Rights, Human Rights: Ecology, Economy, and Ideology in the Canadian Arctic*. Toronto: University of Toronto Press.

Wikipedia 2007. "Kyoto Protocol." Accessed August 7, 2007: Wikipedia.org/wiki/Kyoto_Protocol.

Wilde, Ralph. 2007. Trusteeship Council. In *The Oxford Handbook on the United Nations*, edited by Thomas G. Weiss and Sam Daws. Oxford: Oxford University Press, 149–159.

Yablokov, Alexei V. 1997. On the Soviet Whaling Falsification, 1947–1972. *Whales Alive* 6 (4).

Young, Oran R. 1981a. *Natural Resources and the State: The Political Economy of Resource Management*. Berkeley: University of California Press.

Young, Oran R. 1981b. The Political Economy of the Northern Fur Seal. *Polar Record* 20:407–416.

Young, Oran R. 1982. *Resource Regimes: Natural Resources and Social Institutions*. Berkeley: University of California Press.

Young, Oran R. 1987. The Pribilof Islands: A View from the Periphery. *Anthropologica* 29:149–176.

Young, Oran R. 1989. *International Cooperation: Building Regimes for Natural Resources and the Environment*. Ithaca: Cornell University Press.

Young, Oran R. 1991. Political Leadership and Regime Formation: On the Development of Institutions in International Society. *International Organization* 45:291–309.

Young, Oran R. 1994. *International Governance: Protecting the Environment in a Stateless Society*. Ithaca: Cornell University Press.

Young, Oran R. 1999a. *Governance in World Affairs*. Ithaca: Cornell University Press.

Young, Oran R. 1999b. *The Effectiveness of International Environmental Regimes: Causal Connections and Behavioral Mechanisms*. Cambridge: MIT Press.

Young, Oran R. 2002. *The Institutional Dimensions of Environmental Change: Fit, Interplay, and Scale*. Cambridge: MIT Press.

Young, Oran R. 2005. Governing the Bering Sea Region. In *A Sea Change: The EEZ and Governance Institutions for Living Marine Resources*, edited by Syma Ebbin, Alf Hakon Hoel, and Are Sydnes. Dordrecht: Springer Verlag.

Young, Oran R. 2007. Rights, Rules, and Common Pools: Solving Problems Arising in Human/Environment Relations. *Natural Resources Journal* 47:1–16.

Young, Oran R. 2008. The Architecture of Global Environmental Governance. *Global Environmental Politics* 8:14–32.

Young, Oran R. Forthcoming. Navigating the Sustainability Transition. Essay for volume emerging from the Corsica Workshop.

Young, Oran R., and Gail Osherenko, eds. 1993. *Polar Politics: Creating International Environmental Regimes*. Ithaca: Cornell University Press.

Young, Oran R., Frans Berkhout, Gilberto Gallopin, Marco Janssen, Elinor Ostrom, and Sander van der Leeuw. 2006a. *How Will Globalization Affect Resilience, Vulnerability, and Adaptability of Socio-Ecological Systems at Various Scales?* Global Environmental Change 16: 304–316.

Young, Oran R., Eric F. Lambin, Frank Alcock, Helmut Haberi, Sylvia I. Karlsson, William J. McConnell, Tun Myint, Claudia Pahl-Wostl, Colin Polsky, P. S. Ramakrishnan, Heike Schroeder, Marie Scouvart, and Peter H. Verburg. 2006b. "A Portfolio Approach to Analyzing Complex Human-Environment Interactions: Institutions and Land Use." *Ecology and Society* 11 (2), article 31.

Young, Oran, Gail Osherenko, Julia Ekstrom, Larry B. Crowder, John Ogden, James A. Wilson, John C. Day, Fanny Douvere, Charles N. Ehler, Karen L. McLeod, Benjamin S. Halpern, and Robbin Peach 2007. Solving the Crisis in Ocean Governance: Place-Based Management of Marine Ecosystems. *Environment* 49 (4):20–32.

Young, Oran R., Leslie A. King, and Heike Schroeder, eds. 2008. *Institutions and Environmental Change: Principal Findings, Applications, and Future Directions*. Cambridge: MIT Press.

Young, Oran R., Bradnee Chambers, Joy A. Kim, and Claudia ten Have, eds. 2008. *Biosafety*. Tokyo: UN University Press.

Index